**Countryside Leisure**

**ep** EP PUBLISHING LIMITED
1978

# ASTRONOMY

## PETER CATTERMOLE

EP Publishing Limited
1978

*Dedication*
To my wife who painstakingly read and constructively criticised the manuscript.

Acknowledgements
The author wishes to thank the following for their help in supplying photographs for this book:
H. Dall Esq., pp. 46, 54; D. G. Daniels, p. 14; The Mary Evans Picture Library, pp. 8, 75 right; E. Lindsay, p. 69; NASA, pp. 45 right top and bottom, 52, 56 left, 57, 61; The Science Museum, pp. 11 right, 12 top and bottom right, 13, 48 left; Professor G de Vaucouleurs, p. 55 (both).
Patrick Moore Esq., O.B.E., supplied the cover photograph

**About the Author**
Peter Cattermole, B.Sc., Ph.D., F.R.A.S., F.I.U.P., is a lecturer in Geology at Sheffield University. His professional research includes vulcanicity and the study of lunar and Martian surface features and processes. Having been inspired by Patrick Moore, Peter Cattermole has been observing the Moon and planets as an amateur since 1955 and now has his own observatory.

He has published a number of books including *Exploring Rocks* and *Craters of the Moon* (both with Patrick Moore), *Amateur Geologist, World of Geology* and *All About Space Exploration*.

ISBN 0 7158 0470 7

Published by EP Publishing Ltd, East Ardsley, Wakefield, West Yorkshire, 1978

Text set in 10/11 pt Monophoto Univers by Butler & Tanner Ltd, Frome and London, printed by photolithography and bound in Great Britain by G. Beard & Son Ltd, Brighton, Sussex.

# CONTENTS

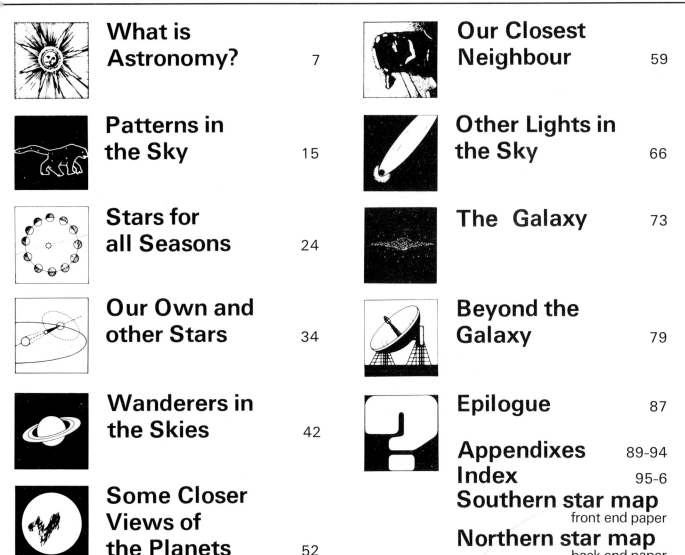

# WHAT IS ASTRONOMY?

Astronomy is the science of the heavens. Early watchers of the skies included the shepherds and the soldiers of early civilisations, who spent much of their time patrolling the hills during the long hours of darkness. Later in history, when men started to venture across the oceans, mariners used the stars to navigate between their own countries and distant ports. Much superstition surrounded study of the stars and planets, and at one time astrologers were paid to forecast the future for their rich patrons.

Today we would laugh at many of the ideas which the early peoples had. This is rather unfair, since we have the advantage of being able to use a storehouse of knowledge that has been accumulated, painstakingly, during the intervening centuries. I wonder, if we had been among their numbers, whether we would have been any more perceptive? I suspect not.

Astronomers of the present use sophisticated and expensive instruments to study the universe. In olden times, the only tools scientists had were their own eyes. Eventually the first small telescopes were made, while today the largest telescopes can search almost unimaginably remote corners of space. Nowadays astronomy does not only progress by using light to penetrate the darkness, but calls on other forms of radiation to assist it. Thus, a modern researcher might be studying some stellar object or other with radio waves, X-rays or infrared light. Astronomy, like other sciences, has progressed further than the early

thinkers could possibly have imagined it would.

Before we survey our subject in the light of modern findings, let us go back in time, and briefly look at some of the early views of the universe. I believe that this will make a fitting start to our journey through space.

### History of Astronomy—Early Theories

Among the earliest of astronomers were the Egyptians, the Chinese and the Babylonians. When the Great Pyramid of Cheops was built, nearly 5,000 years ago, the Egyptians believed the universe to be contained in a large rectangular box. The heavens were supposedly supported on enormous pillars that were set into foundations in a great mountain range. The Sun and the Moon were carried in boats that drifted along on the celestial river, Ur-nes. Egypt was believed to lie in the centre of a flat Earth.

Despite what to us seems a fanciful idea, the Egyptians were by no means fools. Their civilisation was advanced in many ways and their scientific powers were considerable. They used a 365-day calendar which had been established after observations of certain bright stars, in particular the very bright star, Sirius. Furthermore the main internal passage of the Pyramid had been aligned in such a way as to suggest that they had observed the position of the north celestial pole very carefully. The north–south alignment

The Junction of Heaven and Earth. This artistic representation from an unknown source, beautifully depicts one of the early ideas of the universe, where the junction between heaven and earth was a solid dome to which the stars and other objects were attached.

could only have been achieved by observations of the northern stars.

The early Chinese, like the Egyptians, believed in a flat Earth. They, too, had a calendar which was based on simple astronomical observations. The Babylonians supposed that each day the Sun entered through a door set in the eastern side of a solid sky, crossed the heavens, and set through a complementary door in the west. The Babylonian Earth was hollow and inside it was the 'Home of the Dead', where all souls went to rest.

These early notions have since been proved wide

of the mark in terms of reality. Science has developed far beyond the flat Earth–solid sky stage. However, let us not forget that there are still enormous gaps in our present knowledge. Many modern theories may well prove to be quite wrong within five or ten years. This is very often the case with astronomical hypotheses. Fragments of new information come along all the time and throw new light on old theories.

After the decline of the earliest civilisations, it was the rise of learning in Greece which marked the next forward step. Modern astronomy has its foundations in Greek science.

Thales of Miletus (624–547 BC), a noted Greek scientist, although he firmly believed the Earth to be flat, had a burning interest in things astronomical and is said to have gone to the bottom of a deep well to improve his view of the stars. However, while he studied the stars he got no further than supposing them to be fiery jets supported somehow by the 'air'. Another famous Greek, Anaximander, claimed that the Earth was freely suspended in space and although he could not possibly have had the slightest notion of gravity, he must have been the first person to have the idea, however inadvertently.

Pythagoras is sometimes recorded as having been the first to realise the Earth was spherical. In fact Anaximander did so. This insight was not generally accepted, even in the free-thinking Greek culture, and it took Pythagoras, the Newton of ancient times, to reinforce his views before people took the idea at all seriously. It was not until the time of Aristotle (384–322 BC) that astronomical observation produced proof of this fact.

Aristotle made the very valid point that the altitude of the stars changes according to the position of an observer on the Earth's surface. The Pole Star thus appears higher in the sky from Greece than it does from Egypt, while the brilliant southern star, Canopus, rises in Alexandria but is invisible from Athens. Furthermore, Aristotle pointed out that during an eclipse of the Moon the shadow of the Earth appears

curved and so surely the Earth itself must be curved?

Once the idea of a spherical Earth caught on, it became necessary to measure the circumference of the sphere. The Greeks, fortunately, were excellent mathematicians and in about 200 BC, Eratosthenes tackled the problem successfully and came up with an answer very close to the modern figure. His method, a very sound one, relied on the fact that when the Sun is overhead at one place, it is some distance from the vertical at another. Using observations made at both Alexandria and Syene (modern Aswan), he made some relatively simple calculations and came up with a value of 39,760 km (24,850 miles). This is only 80 km (50 miles) less than the modern figure!

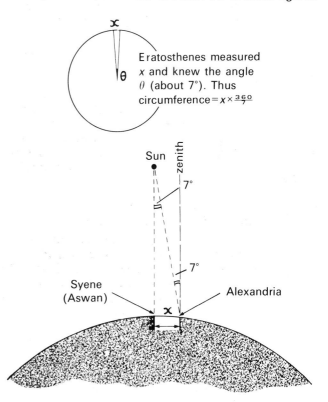

Eratosthenes measured $x$ and knew the angle $\theta$ (about 7°). Thus circumference $= x \times \frac{360}{7}$

The Greek ideas on astronomy were collected together in a work of major importance. Written by Ptolemy, this review was entitled *The Almagest*. It was written during the second century AD in Alexandria where the writer was then librarian. Of great importance though it was, it still contained one incorrect concept which had lingered on over the centuries: it placed the Earth firmly at the centre of the universe. This misconception was to be perpetuated for over 1,000 years more, well after the collapse of Greek civilisation.

The Greek Empire crumbled and the Dark Ages ensued. Culture in Europe fell into a great decline. Eventually the monied rulers of the times decided that their astrologers had no answer to their problems and instructed them to improve the situation. The astrologers, in turn, came to the conclusion that they needed more and better star charts to fulfil their patrons' requests.

Astrologers are fortune-tellers. They seek to predict an individual's future by analysing the position of stars and planets in relation to the subject's birth date. This is supposed to have a decided effect upon both the character and fate of the subject. Astrology is not itself a science but its development at this particular time in history led to the rebirth of its truly scientific counterpart, astronomy.

## Copernicus and His Contemporaries

By the close of the Dark Ages extremely good star charts had been drawn. They were thus available to the contemporary thinkers. Among the new breed of astronomers was a Polish canon, Nicholas Copernicus. A very fine mathematician, Copernicus set about testing the old ideas of Ptolemy and others regarding the movements of the planets. Previously it had been believed that the planets and the Sun moved round the Earth in circular paths, or *orbits*. To explain certain oddities in the planetary movements, however, Ptolemy had to turn to a complex system of 'epi-

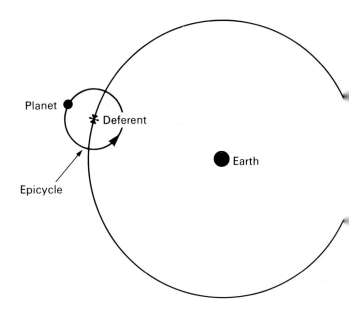

According to Ptolemy the Sun and planets circled the Earth in circular orbits.

cycles'—circles upon circles—to arrive at a reasonably acceptable position.

Copernicus swept aside many of these ideas. In his great book *De Revolutionibus Orbium Celestium* he put the Sun firmly at the centre of the solar system, relegating the Earth to the insignificant role of ordinary planet, along with the others then known, namely: Mercury, Venus, Mars, Jupiter and Saturn.

Needless to say the Church, whose official line at the time was the Ptolemaic view, strongly attacked Copernicus. Being immensely powerful, the Church's stand frightened many potential proponents of the new ideas into silence. They were afraid for their lives to stand out against the ecclesiastical doctrine. Eventually, however, the tide was turned and the Ptolemaic system was cast aside. By the late sixteenth century the famous German astronomer, Johannes Kepler, had worked out that the Earth and other planets must actually move in elliptical orbits.

Theorists had advanced astronomical thinking thus far, but by the middle of the sixteenth century observers of the planets and stars became more numerous. Tycho Brahe, a Danish noble from Copenhagen, made more accurate observations of planetary motions than had been made before. After his death, his assistant, Johannes Kepler, used his calculations to show how the planets actually moved in elliptical orbits, and not circular orbits, as had hitherto been thought. He was to be followed by more observers such as Herschel, Hevelius, Schiaparelli and Schröter. The rise of observational astronomy stemmed from a discovery made by a Dutch spectacle-maker, Hans Lippersheim, in the year 1608.

Lippersheim discovered that by aligning two spectacle lenses, he could magnify the images of distant objects. He had unwittingly invented the first refracting telescope. Such instruments rely on the principle that light, when passed through a lens known as an *object glass*, is refracted and brought to focus near the lower end of the telescope tube. The light rays are then brought and passed through a second lens, or system of lenses—*the eyepiece*—and the image is magnified.

## Galileo

The news of this discovery soon reached across Europe. One recipient of the discovery was Galileo Galilei, then professor of mathematics at Padua University. He heard the news and immediately built himself one of the new 'optic tubes'. Turning it heavenwards, he was amazed by what he saw. The Moon, he discovered, had a cratered surface and was traversed by grey plains from which rose up high mountains. Venus, like the Moon, showed phases, while the planet Jupiter was attended by four satellites. The Milky Way, Galileo found, was made from thousands of individual stars. All this had, hitherto, been hidden from men's eyes.

Galileo was not the only person to make himself a refractor. At much the same time, possibly even earlier, a British observer—Thomas Harriot, one-time tutor to Sir Walter Raleigh—obtained an instrument, probably made on mainland Europe, and viewed the Moon. He even drew a map which appears to have been published as early as 1609, before Galileo's lunar drawing.

More telescopes were constructed and more discoveries made. Slowly the optical components were improved and larger instruments appeared. Galileo's first refractor, which had an object glass a mere 2·5 cm (1 in) across, was a tiny telescope by any standards.

The principle of the refracting telescope. Light, on passing through a lens is brought to focus at a second, smaller lens and the image magnified.

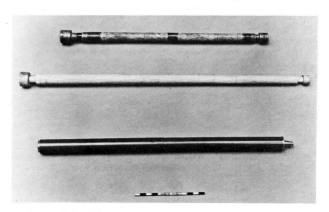

Replicas of the telescopes used by Galileo and Torricelli held in the Science Museum, London.

## Sir Isaac Newton

Later in the same century Sir Isaac Newton, one of the most celebrated scientists of the time, put together a telescope whose optical system consisted of both mirrors and lenses, rather than lenses alone. This was the first reflecting telescope, and it was found to be superior to the refractor in the sense that it produced less aberration of the light which passed through it. The reflector has a main mirror of parabolic form which is placed at the lower end of the tube. This

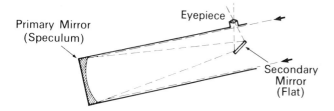

The principle of a Newtonian reflector. Light falls on a parabolic mirror which gathers the light and sends it back up the tube to an inclined secondary mirror (a flat) which deflects to an eyepiece set in the side of the telescope tube.

mirror, or *speculum*, reflects back the incident light so that it arrives at a small, inclined, mirror known as a *flat*. This deflects the light rays towards the side of the tube where they are focused and magnified at the eyepiece. Observing with a reflector, particularly when viewing objects at high altitudes, is much more comfortable in general terms than with a refractor.

It was Newton, inventor of the reflecting telescope, who, in his famous volume *Principia*, outlined the principles of gravity in scientific terms and for the first time provided an explanation of why the planets circled the Sun. Shortly before Newton published his great work, in 1687, the first proper astronomical observatories were opened. One was set up at Copenhagen and another at Leyden; these were the first. The great Paris Observatory was completed in 1671

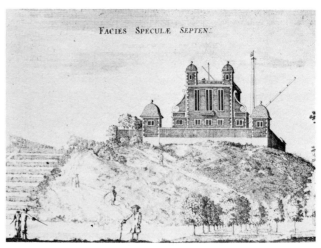

The old Royal Observatory at Greenwich. This old engraving shows the buildings as they must have been in Flamsteed's time.

The long focus refractor telescope of Hevelius, mounted on a tower. Such long focus instruments were very cumbersome to use and fell into disfavour.

Map of the Moon as drawn by Hevelius and published in 1645. This was the best of its time but this worker's system of naming the lunar features was soon superseded by that of Riccioli.

and Greenwich in 1675. King Charles II appointed the first British Astronomer Royal, John Flamsteed, in that year. He was set the task of drawing up accurate star-charts for British mariners, who did all their route-finding by the stars.

Telescopes allowed astronomers to see further and

further into space. More planets were discovered and the surface of the Moon was charted in detail. The Sun's surface was observed by projecting the solar image onto a screen, whereupon the first sunspots were recorded and the Sun was found to rotate on its axis, rather like the Earth. Later still, with the invention of the spectroscope, which analyses the rainbow spectrum produced by splitting white light with a prism, the chemical composition of both the Sun and of distant stars was revealed. Astronomy progressed far quicker than it had ever done before.

## Astronomy Today

Today there are numerous expensively equipped observatories on every continent. The largest reflecting telescope in the world, at the present day, is the great 595 cm (238-in) instrument which has recently been put into operation by the Russians. Not very much smaller is the famous 500 cm (200-in) reflector at Mount Palomar, in California, USA. Then there is the great 100 cm (40-in) refractor at the Yerkes Observatory, USA, and the fine telescopes at Mount Stromlo in Australia. One could go on for several pages listing the excellent instruments now in use.

In addition to visual instruments there are radio telescopes, like that at Jodrell Bank in Cheshire, UK, and at Cambridge University. Some specialised work necessitates the use of X-ray and infrared telescopes, by which means the latest forward steps into the world of 'invisible' astronomy are being taken.

Despite all this sophisticated hardware, astronomy still reveals its wonders to the humble amateur, the armchair scientist and the school student. A pair of eyes is the basic tool of the astronomer. A relatively

The Pleiades cluster. A photograph taken with a small telescope by an amateur astronomer.

small outlay and a little skill can provide the would-be scientist with a small telescope larger than Galileo's, and if this cannot be achieved, a pair of binoculars or a small refractor can be purchased. With either of these tools, a vast world of stars, star clouds, planets and galaxies await the hungry eyes of the observer.

Let us start our celestial journey by taking a look at the main star groups which were first recognised by those early sky watchers of Babylon, Egypt and China.

# PATTERNS IN THE SKY

The Earth is a small rocky world which circles the Sun once every $365\frac{1}{4}$ days. The terrestrial orbit has the form of an ellipse and as it moves along this path it spins rapidly on its axis. One complete spin takes twenty-four hours—one day. As with most planets, the axis is inclined to the plane of its orbit; in the case of the Earth, the angle made between the perpendicular to the orbital plane and the axis of rotation is $23\frac{1}{2}°$. The axis points northwards to what is called the *north celestial pole* and southwards to the south celestial pole.

The celestial pole is an imaginary point in the sky

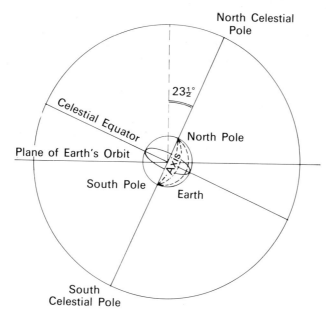

The Earth's axis and its relationship with the terrestrial orbit. As can be seen, the axis is inclined at $23\frac{1}{2}°$ to the orbital plane.

about which the familiar stars appear to revolve. At the present time the position of the north celestial pole is marked by the bright star, Polaris, which is also known as the Pole Star. No prominent star marks the position of its southern counterpart. In the days when the Great Pyramid was being built, this point lay some distance from Polaris, near the star Thuban.

The stars do not in fact revolve about the celestial poles. They appear to do so because of the Earth's rotation. The stars are, to all intents and purposes, fixed, which is why the patterns of stars recognised by early astronomers have remained in use to this day. The patterns they defined represented various mythological characters, for their culture was steeped in legend and myth.

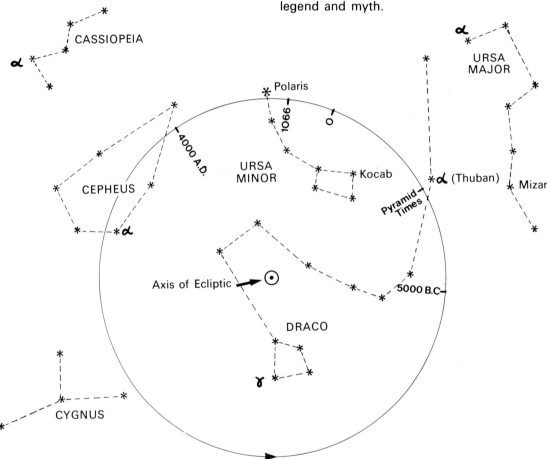

The celestial pole has not always been in the direction of the star Polaris. In the days of the Egyptian dynasties, it lay near the star Thuban in the neighbouring constellation of Draco.

## Constellations—Circumpolar Groups

The early Chinese, the Babylonians and the Egyptians all divided the brighter stars up into groups. The Greeks, however, did the job rather better than these peoples, and it is their groups, or *constellations*, which we use today. Most of them have Greek or Latin names. Thus we have Ursa Major (the Great Bear), Cygnus (the Swan) and Leo (the Lion), while Perseus, Andromeda and Cassiopeia represent mythological royalty.

As civilisation began and developed first in northern latitudes, the southern stars were unknown to these early astronomers. It was not until mariners voyaged southwards in search of riches and new territory, that these stars were first charted. The names they have acquired are thus more recent. The first astronomer to take close interest in them was Sir John Herschel in the nineteenth century.

The most familiar of the northern constellations is Ursa Major, the Great Bear of mythology. The Greeks must be credited with vivid imaginations: the group looks nothing like any kind of animal. This is presumably why many people will know it as 'the Plough' or, if American, 'the Big Dipper'. Call it what you may, Ursa Major has a very distinctive shape and is composed of seven bright stars.

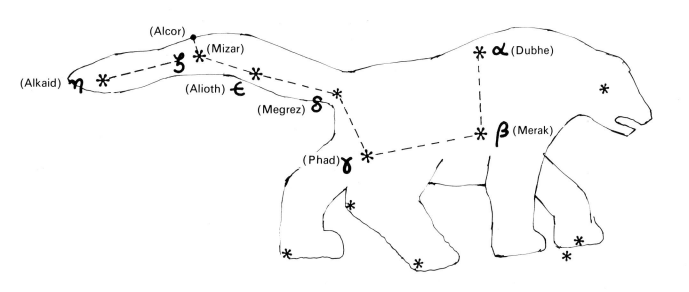

The constellation of Ursa Major, the 'Great Bear' or 'Big Dipper'. It is composed of seven bright stars.

Each of the stars has beside it a Greek symbol, in fact a letter of the Greek alphabet. The policy of allotting each individual star a letter was begun very early on, the idea being to give the brightest star in any group the letter 'A' (Greek: alpha—$\alpha$), the next 'B' (Greek: beta—$\beta$), and so on. As with most plans, the original plan has broken down a little, and we find that alpha of a constellation is not always the brightest star. However, the principle is good and it is a very convenient method of labelling the brighter stars. For those unfamiliar with the Greek letters, I have listed their alphabet below.

### Table 1
### The Greek Alphabet

| | | | |
|---|---|---|---|
| $\alpha$ | Alpha | $\nu$ | Nu |
| $\beta$ | Beta | $\xi$ | Xi |
| $\gamma$ | Gamma | $o$ | Omicron |
| $\delta$ | Delta | $\pi$ | Pi |
| $\varepsilon$ | Epsilon | $\rho$ | Rho |
| $\zeta$ | Zeta | $\sigma$ | Sigma |
| $\eta$ | Eta | $\tau$ | Tau |
| $\theta$ | Theta | $\upsilon$ | Upsilon |
| $\iota$ | Iota | $\phi$ | Phi |
| $\kappa$ | Kappa | $\chi$ | Chi |
| $\lambda$ | Lambda | $\psi$ | Psi |
| $\mu$ | Mu | $\omega$ | Omega |

The brighter stars were also given names by the Greeks. Thus the two stars of Ursa Major known as 'the Pointers'— because they point towards the Pole Star—are called Dubhe and Merak. If one extends a line from Merak through Dubhe, it then reaches Polaris which is in a part of the sky where few bright stars occur. This is the easiest way of finding Polaris and also of locating the direction of north.

From the latitude of southern England, Polaris makes an angle of about 50° with the horizon. As one travels further north, so the angle increases until, in the far north of Scotland, it is 60°. If one was standing at the North Pole, Polaris would be directly overhead; at the Equator it would lie on the horizon.

Ursa Major's seven principal stars form a sort of box with a tail on it. The second star in the 'tail' is called Mizar, or Theta Ursae Majoris, and is rather famous as it has a fainter companion star, Alcor, with which it forms a pair. Mizar is termed a *double star*. There are many in the universe and they are genuine pairs, rotating about one another as the Earth moves round the Sun.

The star in the end of the tail is Eta ($\eta$) and a line drawn from it, through Polaris, brings us to a fairly conspicuous star which forms the corner of a large 'W' grouping. The stars are of comparable brightness to those of the Great Bear and form the constellation known as Cassiopeia.

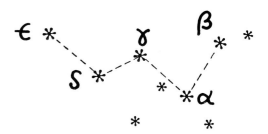

The 'W' shape of Cassiopeia. The second of the familiar circumpolar groups.

In Greek legend Cassiopeia, wife of King Cepheus of Ethiopia, upset the sea god, Neptune, by boasting she was more beautiful than his sea-nymphs. Neptune was enraged and set a particularly obnoxious monster onto her kingdom to terrorise the coast. In their terror the native Ethiopians appealed to their god, Jupiter Ammon, in the desert and he explained that Neptune would be appeased only if Cepheus sacrificed his daughter, Andromeda, to the monster. Eventually he agreed, sadly, and she was duly chained, to a rock. As with many stories, a hero conveniently arrived on the scene. Here the saviour was Perseus, who got rid of the monster, so saving Andromeda and helping Queen Cassiopeia and her husband.

Polaris, Cassiopeia and Ursa Major, once they can be recognised, help in finding other star groups. So, on a convenient clear night, why not go outside and locate the three northern groups? They can be seen from Britain at all times of the year; they never set below the horizon. They are called *circumpolar* stars.

Polaris itself is the principal star in a rather faint constellation, Ursa Minor (the Little Bear). Here there are also seven main stars which form a turned-about version of the Great Bear, bending backwards from Polaris towards Mizar. Polaris apart, the most con-

spicuous star in the group is $\beta$, also known as Kocab.

Not all the stars I have mentioned are of the same brightness. Polaris and Kocab, for instance, are roughly equal but are less bright than either $\alpha$ or $\varepsilon$ Ursae Majoris. On the other hand they are clearly brighter than either $\delta$ Ursae Majoris or $\varepsilon$ Cassiopeia. A star's apparent brightness is known as its *magnitude*.

Ptolemy was the first astronomer to try and list the stars in order of their brightness. He gave the brightest ones a magnitude of one, less bright ones, two, and so on down to the faintest stars visible to the unaided eye, which had a magnitude of six. Since Ptolemy's time, however it has been noted that the stars he listed as of first magnitude are not all of the same brightness.

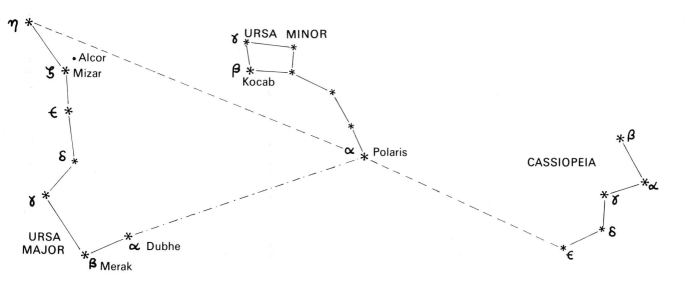

The Pole Star lies in the constellation of Ursa Minor (the 'Little Bear'). The diagram shows its position relative to Cassiopeia and Ursa Major.

Some are very definitely brighter than others. Thus we have the odd situation of certain stars having a negative magnitude! The bright star, Sirius, for instance, has a magnitude of −1·4.

Modern instruments can measure a star's apparent brightness very accurately, so that modern lists are far superior to catalogues like Ptolemy's. I have listed the magnitudes of the three constellations we have just looked at, to give some idea of what the differences really look like. As a guide, there is just one magnitude's difference between $\beta$ and $\varepsilon$ Cassiopeiae.

To return to the star groups themselves: let us now look at a group which can be pinpointed using the stars $\alpha$ and $\beta$ Cassiopeiae as guides. Imagine a line drawn through these two stars and extended for about four times the distance between them. The line reaches a star of the second magnitude which is alpha of the constellation Cepheus. He, you will remember, was the king-husband of Cassiopeia. Cepheus is not a prominent group, but the stars form a recognisable diamond-shape. Two of its stars, $\delta$ and $\mu$, have an

interesting habit of very regularly changing in brightness. They have been of the utmost importance to astronomers and are termed *variable stars*.

Finally, let us not forget the winding, rather faint constellation of Draco, the celestial dragon which is shown on page 16. Its chief star, Thuban, can be found easily by extending an imaginary line through $\gamma$ and $\delta$ Ursae Majoris. The other stars wend their way between Ursa Major and Polaris, curve round towards Cepheus and then bend away again towards the brilliant star, Vega, in the constellation of Lyra.

Those constellations we have so far discussed are all circumpolar. Next we must go further afield and take a brief look at some other groups which are only partly circumpolar or which disappear entirely for a part of each year.

### Partly Circumpolar Groups

One group that is only partly circumpolar from British latitudes is Perseus, the mythological hero who saved the beautiful Andromeda. Its stars are best seen dur-

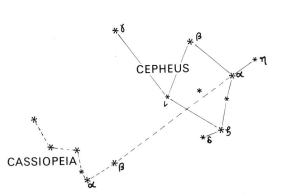

The constellation Cepheus can be found by using two of the stars of Cassiopeia as guides. The interesting variable star, $\mu$, lies in Cepheus.

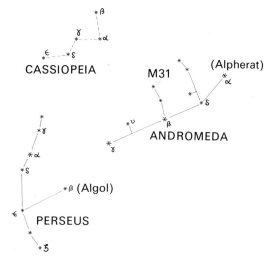

Perseus can be located using $\delta$ and $\gamma$ Cassiopeiae, extending the line about three times the distance between them.

ing winter when all are well above the horizon. To find Perseus, we use the two stars $\delta$ and $\gamma$ Cassiopeiae, extending a line through them for about three times the distance between them. This leads us to a second magnitude star, $\gamma$ Persei. The remaining stars are not particularly striking in terms of their brightness but are distinctive enough as the pattern they make is easy to remember and lies against the backcloth of the Milky Way.

The most famous of Perseus's stars is beta, also known as Algol. This star indulges in some very odd behaviour. It is a variable star and its periodic variations were sufficiently obvious for them to be known to the ancients. It became christened 'the Winking Demon'. The star periodically fades, then brightens up again to a maximum, remaining bright for a period before fading again, and so on. If you watch the star often enough, for long enough, you are almost certain to observe one of its winking fits. It is, in fact, not one but two stars, and is an example of what is called an eclipsing binary. One of the two stars is much brighter than the other, so that when it is hidden by the fainter companion, which rotates about it, the total light we receive is reduced.

Perseus played an important part in the legendary tale I have told. Cassiopeia and Cepheus I have mentioned, and there only remains Andromeda to mention before the story is complete. The diagram below shows how a line drawn from Polaris, through $\beta$ Cassiopeiae and extended for roughly the same distance, brings us to a bright star which sits at the top left-hand corner of a distinct square of stars. The star we have located is called Alpherat and it used to be part of the constellation Pegasus, before astronomers, in their wisdom, decided to move it sideways into the neighbouring group of Andromeda. Since it so obviously forms a part of the box—which is Pegasus— it seems rather nonsensical to have done this, but done it was and nothing can be changed now!

Alpherat is $\alpha$ Andromedae. The remaining stars de-

lineate a curving line towards Perseus, with shorter arms of fainter stars branching off from it. Near to the faint star, $\nu$, which sits at the end of one of these branches, is a faint hazy patch, which is a collection of stars and gas clouds of enormous size, known as a galaxy, and is called M31 Andromedae, or the 'Great Andromeda Galaxy'. It has the distinction of being the most distant celestial object visible to the naked eye, and I shall be saying more about it in a later chapter.

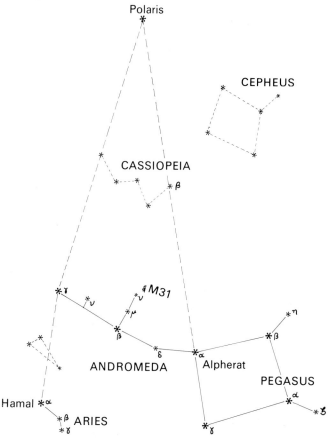

## The Southern Hemisphere

Many of the constellations are 'shared' between observers in the northern hemisphere and those in the south. However, in the same way that the Pole Star can never be seen in Cape Town or Rio de Janeiro, so many 'Southern Stars' are invisible to dwellers in Europe and North America. Thus the rich star fields of Sagittarius are never well-placed for northern observers, and such brilliant stars as Canopus, Alpha Centauri and Achernar are quite invisible.

To observers in high southern latitudes a number of fine constellations are always visible. The figure

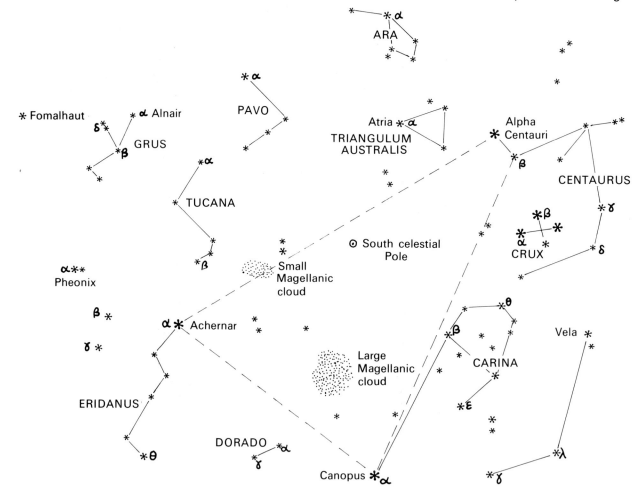

The principal southern constellations and the positions of the two Magellanic Clouds with respect to the Centaurus–Canopus–Achernar triangle.

below shows the principal circumpolar groups, together with adjacent ones. A feature of the south polar region is that it is a rather barren one as far as stars are concerned. No bright star lies anywhere near the pole. The most distinctive constellation is Crux Australis— the Southern Cross. Its four bright stars form a much more compact cross than those of the northern equivalent, the stars of Cygnus. Not far distant are the two principal stars of Centaurus. Alpha Centauri is one of the nearest known stars and is a brilliant whitish colour. The remainder of the constellation is a rather less than distinctive group of fainter stars which form three sides of a box around the Southern Cross.

Slightly closer to the celestial pole than either of the two groups, is the southern triangle—Triangulum Australis. The lead star, Atria, is of second magnitude. On the opposite side of the pole is the brilliant star Achernar, in the constellation of Eridanus, the heavenly river. Achernar and the two lead stars of Centaurus form a large triangle with the second brightest star in the heavens, the brilliant Canopus. The rest of Carina is composed of less bright stars which have an arrangement not unlike a cross. They could sometimes, therefore, be confused with the true Southern Cross, and are thus colloquially termed the 'False Cross'.

Two large star-rich patches lie within the confines of the Canopus—Achernar—Alpha Centauri triangle. These are the Greater and Lesser Magellanic Clouds; each a huge family of stars and gas like our own galaxy. These two spectacular regions are the preserve of the southern astronomer and are particularly well seen from middle latitudes in southern summer.

Having taken a look at some of the circumpolar and partly circumpolar groups and before moving on to look at the seasonal constellations, let us take stock of how the Earth's movements affect our view of the stars at different times of the year.

# STARS FOR ALL SEASONS

Very early in history it was noticed that the planets appeared to move across the sky in a narrow belt occupied by twelve star groups. The belt was called the *zodiac* and the star groups the *zodiacal constellations*. These are the well-known ones listed in horoscopes such as those which appear in glossy magazines.

The Sun's apparent path in the sky is known as the Ecliptic. It crosses the Celestial Equator twice each year, at points A and B.

In those early days, one of the zodiacal groups, Aries (the Ram), held a particularly important place in astronomer's esteem. On 21 March of each year (the Vernal Equinox), the Sun's apparent path in the sky, the *ecliptic*, crossed the Celestial Equator. The exact point where this occurred became known as the 'First Point in Aries'. Today the 'First Point' does not lie in Aries, but in the neighbouring constellation of Pisces (the Fishes); however, Aries is still referred to as the first of the zodiacal groups.

The ecliptic is the Sun's *apparent* path across the sky. Of course, the Sun does not move round the Earth, the situation is quite the reverse. What the ecliptic represents is the trace of the Earth's orbital plane in the heavens. It now becomes a simple matter to see that, this being so, the Earth and the other planets move round the Sun in roughly the same plane.

The Earth orbits the Sun and while doing so rotates on its axis. This is tilted with respect to the ecliptic, such that the Earth's equator makes an angle of about $23\frac{1}{2}°$ with it. It is this tilt which is responsible for our seasons and for the very different views of the stars we get at different times of the year. The stars themselves can be considered 'fixed' in the sky; their positions appear to change simply because of the Earth's motion.

To find the location of a star on any star chart it becomes necessary to know its position; first, with respect to the celestial equator, in exactly the same

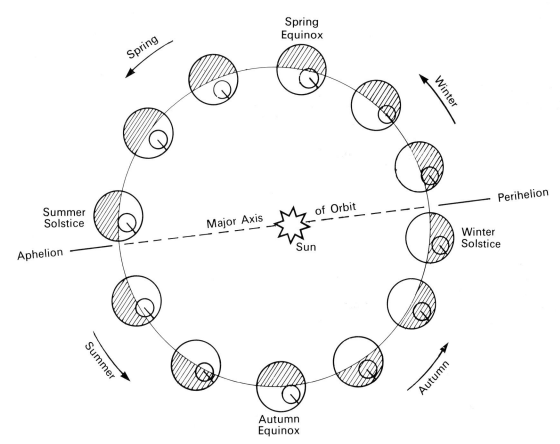

The relationship between the Earth's axis and the ecliptic is responsible for our seasonal view of the stars. The stars themselves can be viewed as 'fixed'. It is the Earth's motion which is responsible for the changes.

way as we need to know the latitude of any town on Earth before we can land an aeroplane near it. Celestial latitude is termed *declination*. Polaris, which is almost at the pole, has a declination of +89·3°. The celestial equator has, of course, a zero declination.

Celestial longitude, known as *right ascension* is the angular distance of a star from the First Point of Aries, and is measured westwards from it. It is given in hours, minutes and seconds of *time*. Time, because each day the First Point of Aries reaches its highest point in the sky. This is called its *culmination*. Any other star will also culminate at its highest point, and the right ascension is the time difference between its time of culmination and that of the First Point. The

right ascension of the next star I shall mention (Capella), is 5 hours 20 minutes, as this is the time interval between its culmination and that of the First Point.

In a way we have been digressing a little from our celestial journey. However, we are now better equipped to deal with certain matters relating to the skies, and, hopefully, this may make these astronomical wanderings much more interesting. Let us first look at the stars of the northern hemisphere, bearing in mind that some of the constellations I mention are visible in the southern hemisphere too, although upside down and at different times of the year.

## Winter Stars

The star which lies overhead during winter is the brilliant yellow star, Capella. This star is a giant version of our own Sun and is the lead star in the kite-shaped constellation of Auriga, the Charioteer. Its position makes it simple to find in winter, and during other times of the year it can be found by imagining a huge triangle with Ursa Major, Cassiopeia and Capella at the corners, as shown below.

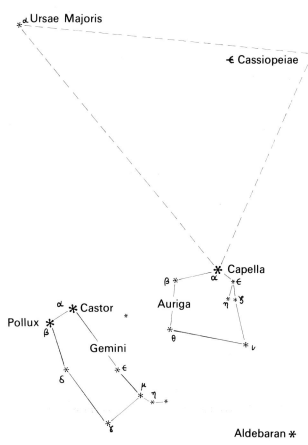

If we continue a line from Capella through β Aurigae we come to two bright stars. One of them, Castor, is white and slightly fainter than its companion, Pollux, which is orange. They are slightly closer together than the Pointers of Ursa Major. Castor and Pollux, the 'Heavenly Twins' of Greek mythology, mark one end of the constellation of Gemini which extends as two lines of fainter stars towards Orion and the Milky Way.

When we looked at the circumpolar constellations, we saw that the Greek legends were for ever imprinted on the heavens via the constellations of Perseus, Cepheus, Andromeda and Cassiopeia. Another legend relates to two star groups which are particularly splendid during the winter months. Both are mentioned in the Bible, for in the Book of Job we find written:

'... Canst thou bind the sweet influences of the Pleiades,
Or loose the bands of Orion?'

The Pleiades were the daughters of Atlas, one of the rebellious Titans who tried to storm the godly stronghold, Olympus. Atlas himself was cast into the pit of Hades and his daughters were pursued by the giant Orion. He, the great Hunter of the skies, still appears to be chasing them across the sky and will do so for ever.

The pattern of Orion cannot be mistaken for any other. It is perhaps the most spectacular of the winter groups and can be seen in the southern part of the sky during a clear evening. If in doubt, extend the bottom row of stars in Gemini towards the bright orange star, Betelgeuse (pronounced Beetlejuice) which marks the Hunter's head.

Betelgeuse, or α Orionis, is a giant orange star and makes a fine contrast with the brilliant white star, Rigel, that forms the Hunter's feet. Rigel, or β Orionis, is the seventh brightest star in our skies even though it is incredibly remote. It beams out across the Universe like a giant searchlight with a power equal to

50,000 suns.

The rest of Orion is also distinctive. The belt is picked out by the stars, $\zeta$, $\varepsilon$ and $\delta$ and the stars of the sword spread down below these three. If you were gazing up at Orion you could not help but notice a brilliant bluish-white star to the left of it and slightly nearer to the horizon. This is Sirius, the 'Dog Star' and it is the most brilliant of all the stars with a magnitude of $-1\cdot4$.

Sirius marks the head of Canis Major, the 'Greater Dog' who accompanied Orion on his hunting trips. It is a very hot star and is also relatively close to the Sun which is why it appears so bright in the sky. Another very hot, close star forming a triangle with Sirius and Betelgeuse is Procyon, chief star of Canis Minor (the 'Lesser Dog'), which is easy to spot as it lies in a part of the sky where there are few other bright stars.

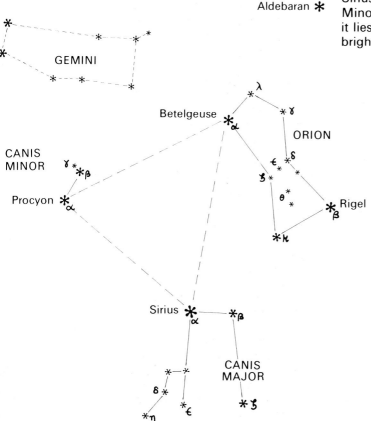

The heavenly hunter, Orion and his two hunting dogs—Canis Major and Canis Minor. These groups dominate the southern sky in the winter months.

If we return to Orion for a moment, and draw an imaginary line from Betelgeuse through $\lambda$ Orionis, we come to an orange star called Aldebaran which is $\alpha$ of the constellation of Taurus (the Bull). Aldebaran has a magnitude of 0·9 and there is a triangular group of stars near it which includes what is called the Hyades Cluster, a very rich star-studded part of the sky. This is well worth looking at through binoculars.

Lastly in our tour of the winter sky, we must mention the very prominent group of faint white stars called the Pleiades, or 'Seven Sisters' which lies a little to the right of a line between Aldebaran and Algol. At first sight it appears as a fuzzy patch, but a closer look reveals it to be a group of seven or more faint stars. This cluster of stars is a fine sight in binoculars. Try counting the stars you can see!

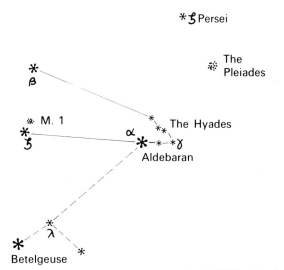

Betelgeuse can be used to locate Taurus, the Bull. The Hyades cluster sits within this group and it makes an interesting binocular sight.

## Spring and Summer Stars

As spring comes along and the nights start to get shorter, Ursa Major is almost overhead. If we follow the bear's tail (or the handle of the plough, if you prefer to think of it this way), towards the horizon, we arrive at the dazzling Arcturus, second only to Sirius in our skies, and $\alpha$ of the constellation of Boötes, the Herdsman.

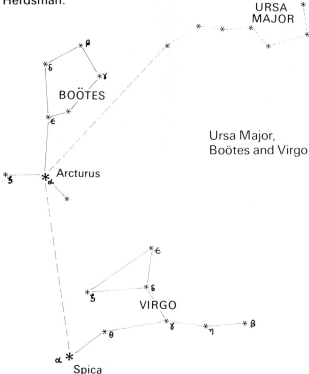

Ursa Major, Boötes and Virgo

The rest of Boötes is rather insignificant, so let us continue a curving imaginary line even nearer to the horizon, when we shall come upon Spica, the brilliant white star in the constellation of Virgo (the Virgin).

Spica is always rather near to the horizon and is therefore usually seen through a layer of haze. This means that its pure white colour is reddened somewhat, which is a pity as white is the colour associated with purity. Spica, or α Virginis, has a magnitude of precisely 1.0.

In the southern part of the sky, the constellation of Leo (the Lion) is dominant. This group really does

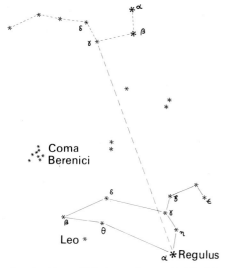

The stars δ and γ Ursae Majoris can be used to find Leo, the Lion. The chief star of this group is Regulus.

resemble its name, the Lion clearly sitting on his haunches with his head up in the air. As can be seen from the chart the chief star is Regulus (α Leonis), and this can be found easily from Ursa Major by using the stars δ and γ.

Regulus sits at the lower end of a line of stars that form the lion's head and front feet. Because they look like a reversed questionmark, they are also known as 'The Sickle' on account of their shape. The lion's hindquarters are formed by a triangle of stars that includes β. This is an interesting star because the ancient scientists considered that it was as bright as Regulus. Today it is certainly fainter, and so we may be looking at a star which has actually faded over the centuries.

Towards midsummer, Capella will sink towards the northern horizon and be barely visible if it is at all misty. Pride of place in the overhead position then goes to the brilliant bluish-white star, Vega, chief star in the constellation of Lyra (the Lyre). Vega dominates its companion stars which are rather faint, and only Sirius and Arcturus are brighter.

Not far away and forming a very acute triangle with Vega and Cassiopeia, is Deneb. This lies at the head of Cygnus (the Swan) and is of magnitude 1·3. The stars of Cygnus look nothing like a bird, being much more like a cross; this is why many people prefer to call it the 'Northern Cross'.

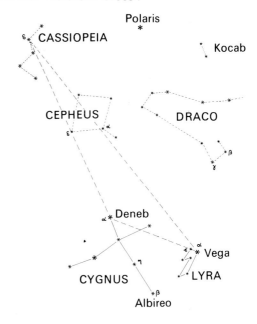

Overhead in midsummer is the brilliant star, Vega, in the constellation of Lyra. Close-by is the cross-shaped group, Cygnus and the less well-defined constellation of Aquila.

Deneb, like Rigel, is another celestial searchlight. It is enormously luminous, but very remote indeed, and the light we receive from Deneb left that star while the Romans still ruled Britain! Astronomers have calculated that it shines with the strength of 10,000 suns.

The star which forms the foot of the cross is called Albireo and it is one of the stars worth looking at with a telescope. It is a beautiful double star, the two components being of contrasting colours.

Deneb and Vega form a triangle with Altair, or α Aquilae. Aquila (or the Eagle) is otherwise not particularly bright, but spreads across the Milky Way and is fairly easy to find.

Last among the summer stars I want to mention is a bright orange-red star called Antares. You may be lucky enough to spot it low down on the southern horizon in summer if you are away from street lighting. The star group of which it is the lead star is Scorpio (the Scorpion) and it lies well south of the Equator. Antares itself can be found by extending a long line from Deneb, through Vega towards the horizon. It is a very distant star and is of comparable size to Betelgeuse, the other orange-red star in Orion.

### Autumn Stars

By the time the evenings start to draw in during autumn, the appearance of the sky will again have changed. Ursa Major is now low down in the north,

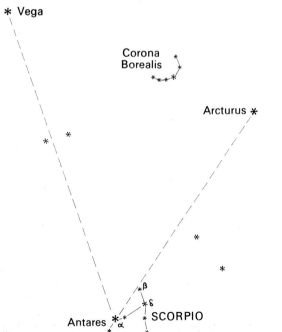

The huge red giant star, Antares, is found in the constellation of Scorpio. This may be found by using the Great Bear as a pointer.

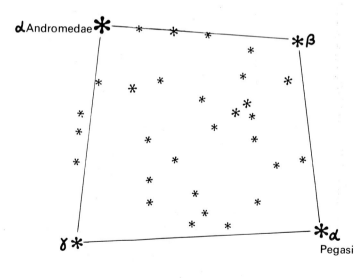

A naked-eye view of the square of Pegasus, drawn by the author on October 15, 1955. Most of the small crosses represent stars whose magnitude is five or lower.

while Deneb and Altair are quite prominent, the former lying almost overhead. Cassiopeia too, is high up and can be used to pick out a constellation we have not yet come across. This is Pegasus, the 'Flying Horse' which now dominates the southern sky.

The 'Square of Pegasus' is clear enough to see although the bottom two corners may be rather faint if there is a mist near the horizon; α Pegasi is the bottom right-hand star of the square and it has a magnitude of 2·5. Within the square there are few bright stars, and it may be of interest to see how many can be found, and to chart them.

Stretching away from the top left-hand corner of Pegasus is a line of bright stars that forms Andromeda, the legendary daughter of Cassiopeia. The line of stars eventually joins Perseus, which is fitting since it was he who saved Andromeda and Cassiopeia from Neptune's monster.

The southern horizon at this time of the year is rather barren. However, you may be fortunate enough to spot the bright star, Fomalhaut, chief star in the constellation of the Southern Fishes. It is best observed about 8 o'clock on an October evening, when it is just skimming the horizon.

Last of the autumn groups I am going to mention is Aries (the Ram). Its chief star is Hamal and this star forms a quadrilateral with γ and α Andromedae and γ Pegasi. See page 30.

We have arrived at the end of our journey through the northern skies, and it is very fitting that we should finish by finding Aries because of its special importance first noticed by the ancient astronomers.

Let us now look at the appearance of the southern sky through the different seasons, assuming the observer to live on or around latitude 26° south. This will thus be applicable to dwellers in South Africa, South America and Australia.

During the summer months the brilliant Achernar is overhead, while Crux Australis is very low in the sky. The two Magellanic clouds are very prominent as also is Orion and the superb Sirius, in Canis Major.

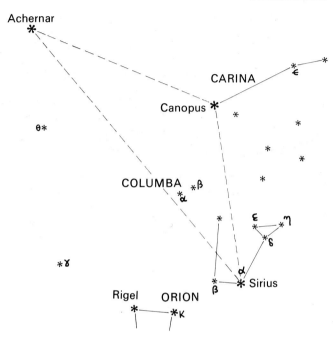

During summer the brilliant Achernar is overhead. An imaginary triangle joins this star with Canopus in Carina, and the brightest star in the heavens, Sirius, in Canis Major.

Canopus, too, is high in the heavens by late evening. The lonely bright star Fomalhaut, always so low down for northern observers, is a fine sight also.

By late February the Southern Cross is becoming more prominent in the south-east, as too are Alpha and Beta Centauri. Fomalhaut by this time will have sunk in the south-west, but Gemini rises gradually in the north-east, on the opposite side of the sky dome. By March Canopus is at the zenith and excellent rich starfields will be worth sweeping with binoculars in the constellation of Carina. During the autumn months Crux rises higher and higher in the

31

sky and the Milky Way, passing through Crux, Argo, Orion and Auriga is a fine sight. Achernar by now will have descended in the south-west but later in the evenings Leo and Ursa Major become visible.

As the evenings draw in, so the rich groups, Centaurus and Crux become prominent. Scorpio is a superb sight in the south-east, the brilliant orange giant, Antares, being alpha of this group. By now

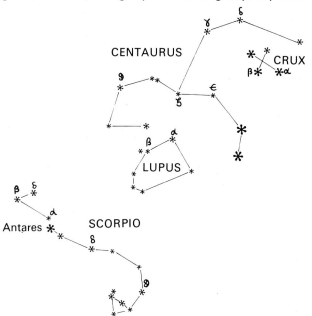

Autumn groups: The Southern Cross (Crux) and Centaurus are prominent and are separated from the winding constellation Scorpio by the fainter stars of Lupus (the Wolf).

Sirius has sunk below the horizon but in its place will have risen the fine star, Arcturus, which can be located in the northern part of the sky. Even the Northern Cross is now to be seen. In the north also are the less than distinguished groups of Hercules and Ophiuchus, although α Ophiuchi is easy to spot (see right). Sagittarius becomes prominent and while the group itself is not easy to define, there can

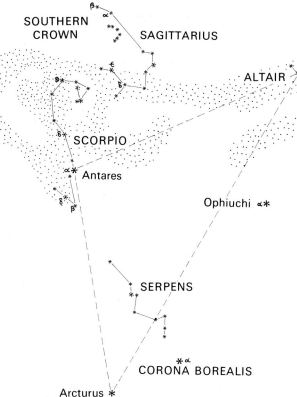

Scorpio lies amongst rich star fields, as does the less conspicuous group, Sagittarius. An imaginary triangle joins the reddish star Antares to the brilliant Arcturus, seen in the north, and Altair, chief star of Aquila, another constellation shared between the hemispheres.

be no mistaking the rich starfields amongst which it lies. Binoculars will reveal an enormous number of stars in this region.

Virgo (the Virgin), a constellation familiar to northern observers, is quite high at this time and as winter moves on it remains so. By August Scorpio occupies the zenith, with Centaurus and Crux high too. However, at this time of the year Canopus is at its lowest. The northern sky is rather dreary, being taken up largely by the unspectacular groups of Hercules and Ophiuchus.

In springtime Fomalhaut rises high in the south-east, while Sagittarius and the rich starfields are overhead. Achernar starts to rise again, and can be seen in the south-eastern part of the sky. In the north the constellation of Aquila, with Altair ($\alpha$ Aquilae) can be seen and Vega is lower down in the west. On a good night it should be possible to make out Deneb very low down in the north-east, and Cygnus too, although the latter is always very low.

The overhead position is occupied by the 'Southern Birds' in October. None of these groups has any very bright stars but the constellation, Grus (the Crane) is easy to spot. Both of its chief stars are of magnitude two. Alnair ($\alpha$) is white and $\beta$ Gruis is a warm orange hue. They make a lovely pair. The other birds are Tucana (the Toucan) which contains the bright globular cluster 47 Tucanae, Phoenix (the Phoenix) and Pavo (the Peacock).

November sees Canopus rising in the south-east and Fomalhaut at the zenith. The Southern Birds remain prominent and Achernar is by now high too. The two Magellanic Clouds are a fine sight while the square of Pegasus begins to be visible low down in the northern sky. Some of the stars of Andromeda can be seen leading away from it. By Christmas Achernar is close to the zenith and the winding constellation of Cetus (the Whale) high in the sky.

One thing which is undeniably strange to a northern observer, is seeing Orion 'upside-down' in the sky. This of course is the case when one travels to southern climes. It would be equally odd to a southern observer to find that Sirius is actually lower in the sky than Betelgeuse when seen from the latitude of London or Prague. Readers of this book in different parts of the world may find a little judicious turning of the pages of star maps into different orientations, will soon get them to 'fit' their own particular situation.

# OUR OWN AND OTHER STARS

## The Sun

The Sun is composed entirely of gases, of which hydrogen is by far the most abundant. Deep inside the star, which has a diameter of 700,000 km (434,700 miles), the gas hydrogen is being changed into another gas, helium. Every time this reaction occurs the Sun loses a little of its mass but gives out energy. We feel this energy as heat and see it as light. If the Sun did not work in this way, life on Earth could not exist.

The process of which hydrogen is converted to helium is known as a *nuclear reaction*, and is similar to that which occurs when an atomic bomb is exploded. However, the amount of energy released by the Sun each second is many times greater than all the bombs Man has ever exploded.

Over one million Earths could be packed inside the Sun but only 332,000 Earths would be needed to balance the Sun if the two worlds were put on the pans of a balance. This means that the atoms of gas which make up the Sun are less closely packed than those which build the Earth. In other words, the Earth is denser than the Sun.

The Earth is, of course, a planet. There are eight other planetary worlds and all of these are part of the Sun's family. The smaller of the planets, including Earth, are rocky and dense, while the larger ones are built from ice and gases which are very much compressed. All are much denser than the Sun, but are

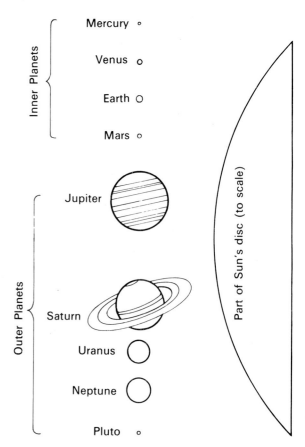

The sizes of the Sun and its family of planets. The smaller inner planets are quite dense, far more so than the large outer group.

less massive because they are very much smaller.

*The Sun should never be looked at through either binoculars or a telescope. To do so will result in almost certain blindness.* To view the Sun safely, either get hold of several exposed sheets of negative film and look at it through these, or project the Sun's telescopic image onto a white card held some distance from the eyepiece.

The bright surface of the Sun which we see daily and which is seen by projecting its image, is known as the *photosphere*. It has a temperature of about 6,000 °C. From time to time, the bright photosphere is dappled by darker patches known as 'sunspots'. These are large areas of the surface which are cooler than the rest, and therefore give out less light and heat. Very large spots can sometimes be seen with the naked eye if viewed through darkened film and even the smallest are far larger than the Earth. A day-to-day survey of a prominent sunspot group reveals that it traverses the solar disc from east to west as the Sun spins on its axis. A complete rotation takes about 27 days—the Sun's *synodic period*.

We cannot see directly inside the Sun, but it is known that the Sun's central core is the real store-house of energy. It is here that the nuclear reactions take place which release the energy that makes the Sun 'shine'. In giving out this energy our star loses mass at the rate of four million tonnes per second. So, since last week, for example, it has lost the almost unbelievable amount of 2,500,000,000,000 tonnes.

It is hard to imagine that the Sun can lose mass at this rate without suffering some damage. However, it has been in existence for over 4,000 million years, and its fuel supply is unlikely to be used up for an even longer period in the future. There is no reason to worry about the Sun 'going out'.

When the Greeks studied the movements of the Earth, Moon and Sun, they found that the orbits of the three worlds lay almost in the same plane. So, not surprisingly, from time to time the three bodies were perfectly lined up, and when this occurred the

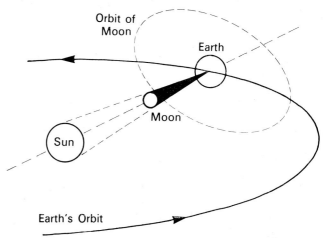

Diagram illustrating how a solar eclipse occurs. Although the Moon is very much smaller than either of the other bodies, it is very near to Earth and is of just the right size to eclipse the Sun when the alignment is perfect.

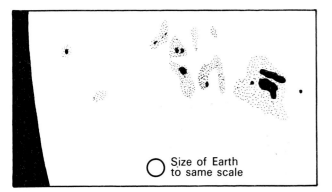

Size of Earth to same scale

The large sunspot group of 1959, as drawn by the author by projection through a 3-in refractor.

Moon eclipsed the Sun entirely and gave a *total eclipse*. Of course, the Moon is very tiny when compared to the Sun. However, it is very much nearer to the Earth than it is to the Sun and looks the same size as the Sun in the sky. Hence it is able to obscure the solar disc during eclipses.

During a total eclipse we are able to glimpse, for a brief moment, the outer part of the Sun's atmosphere. Normally this cannot be seen, as the bright photosphere drowns it, but for a few minutes the glorious *corona* and also the *chromosphere* are seen and can be photographed.

The chromosphere rises for about 10,000 km (6,210 miles) above the photosphere. In this layer the solar gases are in very rapid movement and a great deal of energy is emitted, the temperature being about one million degrees centigrade. The total number of gas atoms present is, however, very small so that the total amount of heat given out is not as great as might be expected.

The solar corona during the 1962 total eclipse.

Higher still above the surface there is the corona. This is the outermost part of the Sun, where the atoms of gas are very far apart and gradually get fewer and fewer until the Sun merges into space. The high level particles move above at very high speed and from time to time enormous flare-ups of hot gas spread up into the corona from the bright photosphere. The most violent of these give birth to *solar flares* which send

out radiations that may stretch out as far as the Earth's orbit.

Solar flares also send out radio waves. The waves of energy here have a much greater *wavelength* than light waves which are shortwave radiations. These longwave radiations reach Earth and can be collected by radio telescopes—instruments which astronomers only began to use quite recently. Radio astronomy is an important new branch of the science and the study of waves coming from our own star has helped in our understanding of the very distant stars (see chapter 10).

Our knowledge of the Sun's chemical composition depended upon the important discovery made by Isaac Newton over 300 years ago. Newton discovered that a glass prism split ordinary 'white' light into

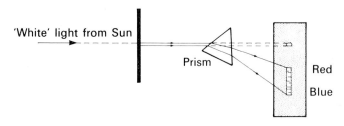

Isaac Newton discovered that white light could be split into its constituent 'rainbow colours' by using a prism. This discovery led astronomers to an understanding of the composition of stars, including the Sun.

its individual colours—the colours of the rainbow in fact. The rainbow colours he observed form what is called the spectrum. About 150 years after Newton had conducted his prism experiment, a German optician called Josef Fraunhofer passed the Sun's light through a prism set in his astronomical telescope, and thus obtained a solar spectrum for the very first time. He was surprised to find that in addition to the rainbow effect found by Newton, there were also some distinctive dark lines crossing it. These became known as Fraunhofer lines.

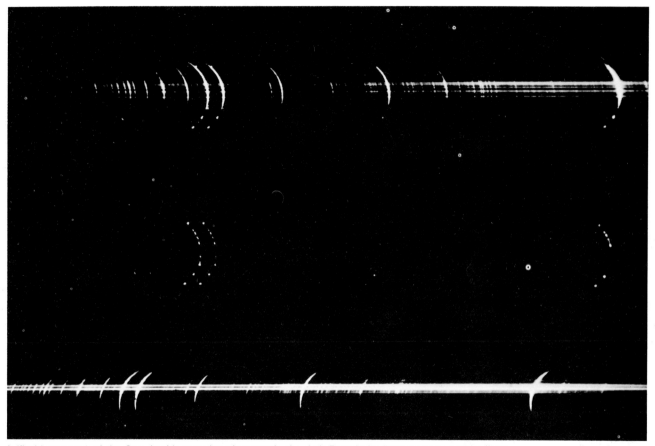

A 'flash' spectrum of the Sun. In this negative photograph, the dark Fraunhofer lines appear as white lines.

To Fraunhofer these lines were a mystery but as time went by it was realised that they were due to the presence of common elements in the Sun like sodium. Since the lines for any one element always occurred in the same position within the spectrum, it eventually became possible to investigate the gases within the individual stars, even though they were extremely distant.

Gradually better and better spectra were obtained as the science of *spectroscopy* made headway. Over seventy of the ninety-two known natural elements have now been found in the Sun. In fact the gas helium was found on the Sun before it was found on Earth!

Astronomers now have a very good idea of how the stars differ from each other. We have seen how they are of varying colours during our survey of the constellations. Star colour is a reflection of heat in the same way as are the various parts of a flame. The bluish cone in a flame is the hottest part, the yellow zone is somewhat cooler and the outer (red) zone is cooler again. So it is with the stars. Bluish stars, like Spica, are extremely hot and have a surface temperature of around 28,000 °C. White stars like Sirius are a little cooler, while yellow stars such as the Sun and Capella have surface temperatures of about 6,000 °C. The cooler stars are the orange and red ones, like Betelgeuse and Aldebaran, which have surface temperatures of about 3,500 °C.

At the close of the last century a well-known pair of American observers, Fleming and Pickering, made a close study of the spectra of thousands of stars and drew up a classification based on the spectra they found. They suggested that there were eleven distinct types of spectra and so divided all the stars into one of these *spectral classes*. I have shown the various classes in Table 2 and have tried to illustrate how the spectra, temperature and chemical composition of the different types are related and to which of the groups some well-known stars belong. Our own star, you will see, belongs to class G.

Despite its losing mass at very rapid rate as it pro-duces energy, the Sun is a stable star and will be able to continue behaving like this for thousands of millions of years to come before starting to become senile. Other stars are not so fortunate. Rigel, the brilliant white star at the feet of Orion, is one such example. It is an amazingly energetic star and gives out 50,000 times the Sun's total energy output. You can imagine, I am sure, that if it is doing this it must also be losing mass at an enormous rate. In fact it loses it to the extent of 80,000,000,000 tonnes each second! Even the very large star, Rigel, cannot continue to do this for long, so it must age very quickly and will die long before the Sun does.

### The Life-pattern of Stars

What then, is the life-pattern of a star? Do stars have a birth and a death?

In the sword of Orion, near to the star $\theta$ Orionis, it is possible to pick out a faint hazy patch of light. If we were able to look at this object through a very large telescope, we should see that it was a huge cloud of illuminated gas. Such a cloud is called a *nebula* and this one is the 'Great Nebula in Orion'.

There are many nebulae in the universe and it

### Table 2
### Spectral Types and Examples of Naked Eye Stars

| Star class | Temperature | Colour | Elements prominent in spectra | Examples |
|---|---|---|---|---|
| W and O | Rare stars and none prominent | | | |
| B | 28,600 °C | Bluish | Helium | Spica |
| A | 10,700 °C | White | Hydrogen | Sirius |
| F | 7,500 °C | Yellowish | Calcium | Cassiopeia |
| G | 5,000–6,000 °C | Yellow | Various metals | Sun, Capella |
| K | 3,000–5,000 °C | Orange | Hydrocarbons | Arcturus |
| M | 3,400 °C | Orange-red | Calcium and titanium oxide | Betelgeuse |
| R, N, S | less than 3,000 °C | Red | Carbon | U Cygni |

seems that these are the birthplace of stars. For some reason, molecules of dust and gas which are scattered through space, gradually come together and congregate. Slowly, more and more particles move towards the focus until the force of gravity starts to draw in even more matter. In a way it works in a similar fashion to bees swarming onto a honeypot, but in the case of our gas cloud it is the force of gravity and not honey which draws the material together.

M42 Orionis: the great nebula in Orion. This huge cloud of dust and gas appears as a hazy patch near to θ Orionis.

Over millions of years, gravity attracts sufficient dust and gas for quite a dense cloud to form. This then starts to become hot at the centre and we have thus arrived at the beginning of a new star's life. Once heat is generated in the cloud, energy starts to be given out and nuclear reactions, like those taking place deep within the Sun, commence. Hydrogen is the main fuel of stars, so it is not surprising that this gas is abundant in the Orion Nebula and most others, and can be picked out in the spectra of these clouds.

Once the young star starts to generate heat, it starts to shrink rapidly, becoming denser all the time. How long this first stage takes depends on the size of the original cloud of matter. In the case of our Sun, which

probably formed from a moderate-sized cloud, this first stage may have taken about thirty million years. A more massive star would take less time, a less massive one longer.

Eventually the shrinking slows right down, and nuclear processes start to convert the hydrogen gas into helium, giving out radiant energy. This marks the beginning of the star's adult life. The Sun is some way through this part of its life at the moment.

Slowly, over hundreds of millions of years, more and more of the hydrogen fuel supply is used up and the star loses more of its mass until, at some critical stage, a change starts to occur within it. This probably happens when the star has used up all of its hydrogen and is left with a core of helium. Helium is what is known as an *inert* gas. This means it cannot take part in chemical reactions as hydrogen can, so the star has to find other ways of generating energy.

Stars similar in size to the Sun do this in a very odd way. What happens is that the outer layers start to expand so that the very dense central core of the star becomes surrounded by a thick envelope of rarified gas. The core will still be very hot but the outer envelope much cooler, perhaps only 3,000–4,000 °C. Old stars which have reached this stage in their life are called 'Red Giants'. Betelgeuse is one such star and another is Antares, in the constellation of Scorpio. Both are very large cool orange-red stars.

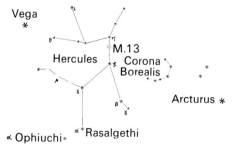

The constellation of Hercules, a rather faint group which can be found by using the stars Arcturus and Vega. Rasalgethi lies within this group.

Even larger is another star, this one lying in the rather faint constellation of Hercules, which can be found from Vega and Arcturus. The principal star in this group is called Rasalgethi and it is of the third magnitude; so it is not a spectacular object in the sky by any standards. Its surface temperature appears to be about 2,700 °C, and it belongs to spectral class M. However, its diameter is enormous and has been estimated at 400 million km (250 million miles), which is greater than the diameter of the Earth's orbit round the Sun. It is a true giant in all senses of the word!

We are less sure of what happens once a star has become a Red Giant but what we can be sure of is that this stage cannot last for ever. Sooner or later, all the available fuel is used up and it changes its structure again. This time it shrinks to an incredibly dense white body known as a 'White Dwarf'. White Dwarfs are all very faint and none can be seen with the naked eye. The bright star, Sirius, has a White Dwarf companion, but we cannot see it. These feeble old stars give out little light but they are so dense that a marble made from part of one would weigh many tonnes! It is almost impossible for us to imagine anything as dense as this.

A star, then, starts its life in a cloud of dust and gas like the Orion Nebula. Gradually it shrinks and becomes more stable. Nuclear energy is generated within it and it becomes an adult like the Sun or Capella. After a very long time the energy store is used up, changes take place and the star expands into a Red Giant, like Betelgeuse or Rasalgethi. Finally, when it is quite bankrupt, it shrinks to become a dense White Dwarf and finally may 'go out' altogether.

At the present time the Sun is in its adulthood. Several millions of years from now, it will look the same as it does today. However, when its fuel supply is used up, it will probably slowly expand into a Red Giant. This will have very serious effects on our planet because, although the Sun's surface temperature at that stage would be lower than at present, the total amount of energy it will be giving out will be much greater. The oceans will boil and the atmosphere will be whisked away. Life on Earth will perish.

Stars such as the Sun, Vega and Betelgeuse are what we might call 'normal' stars. Not all stars are like this. Some change their brightness either very regularly or quite irregularly. Such stars are known as *variable stars*, and there are very many of them. Variable stars have set astronomers a very difficult task because somehow we need to know why the changes they go through take place. We have a good idea why some variables act the way they do and I shall describe one or two in a later chapter. However, many are very puzzling and need to be observed as often as possible, even by the amateur astronomer.

So far I have said little about the distance of the stars. The Sun, we know, is 150 million km (90 million miles) from Earth, but what about the more distant stars?

First of all, let us be clear about what the Sun's distance means; the figure itself does not really mean much, and the best way to try and get some meaning into such a large figure is to construct a scale model. Let us, therefore, shrink the Sun to a globe 1 m (3 feet 3 in) across. On this scale, the Earth would need to be a marble, and would sit 100 m (330 feet) away. The Moon becomes a ball-bearing 3·0 mm (⅛ in) in size and has to be put a quarter of a metre from the marble. The nearest of the stars, if we wished to include it in our model, would be found 20,000 km (124,000 miles) away. Thus if we placed our model Sun at one end of a soccer pitch, the Earth would be found at the other and the nearest star in Australia!

The nearest star, in fact, is in one of the southern constellations, Centaurus. The star itself is called Proxima Centauri.

Light travelling at a speed of 300,000 km per second (186,000 miles per second) takes only eight minutes to reach us from the Sun, a few seconds from the Moon, but over four years to reach us from Prox-

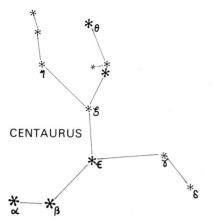

CENTAURUS

The nearest known star, Proxima Centauri lies in the southern constellation of Centaurus. Light from Proxima takes about four years to reach Earth.

ima. Astronomers say that Proxima Centauri is just over four *light years* away.

A very long multiplication would tell us that light, travelling at the enormous speed it does, would cross nearly 95 million million km (60 million million miles) of space in one year. I will leave the reader to work out how far away Proxima is! What it does show is that stars are a very long way from us, even the nearer ones, and that both km and miles are useless as units of measurement. This is why astronomers use the *light year* as their unit of distance.

Most of the stars are much further away than Proxima Centauri. Here are some examples from the constellation of the Great Bear: the star $\varepsilon$ is 68 light years away, $\alpha$ is 107, and $\eta$ is 210, which is as convenient a point as any to illustrate that the stars we see as 'the Great Bear' are really not related to each other at all; they are at quite different distances from us. Most of the constellation patterns we see are simply 'line of sight' effects of this type. This may come as a disappointment, but it is true.

What is also interesting is that stars of one group which may be of the same *apparent* brightness may, in fact, be of quite different absolute brightness because some are much nearer than others. Just look at the case of Ursa Major, where the star at the end of the bear's 'tail' ($\eta$) is over three times as distant as Dubhe ($\alpha$). Clearly it must be very much more luminous than Dubhe if it appears to be of the same magnitude. Then again, Polaris (the Pole Star) does not itself seem very striking. It is only of second magnitude and so is slightly fainter than some of the stars in Ursa Major. However, Polaris lies at a distance of 680 light years and in fact is equal to 3,000 suns. If we were able to view the Sun from Polaris, it would be a *very* inconspicuous object in the sky. The universe exists on a very grand scale.

# WANDERERS IN THE SKIES

At least 5,000 years ago it had been recognised that five 'stars' did not remain in fixed positions like the others. These were called the planets, and were christened Mercury, Venus, Mars, Jupiter and Saturn. The Greeks subsequently plotted their movements, but it was not until the seventeenth century that it was generally appreciated that they moved round the Sun in elliptical paths.

Once the telescope had been invented and pointed towards the heavens, a new era had dawned in astronomy. Eventually, in the eighteenth century a seventh planet was discovered, and later an eighth. Since that time only one other has been found, the diminutive world Pluto, that skirts round the fringes of the solar system at an immense distance and which the early astronomers had no chance of seeing.

## Venus

Almost certainly the first planet to be seen was the most striking of them all, Venus. It was given the name of the goddess of love because of its pure white colour, traditionally associated with virginity. At first it was believed that there were two brilliant white planets: one, known as the 'Morning Star' could be seen only close to dawn, and the second, or 'Evening Star', was only visible in the skies around sunset. Eventually, it was realised that these were one and the same body, but seen at different times of the year.

Venus is seen only close to the time of either sunrise or sunset because it is much closer to the Sun than Earth, and follows our star very closely in its wake. It can be seen in broad daylight, with a telescope, if one knows exactly where to look. Venus is so bright, that using a star chart and a setting device on a telescope it is possible to see it during the daylight hours, despite the brilliance of the sky background.

Its brilliance is due to its surface being shrouded in highly reflective clouds. These reflect nearly sixty per cent of the sunlight that falls on them. Thus we say Venus has an *albedo* of sixty per cent, compared to the mere seven per cent of the Moon. You can see why Venus is such a brilliant object.

Originally Venus was thought to be slightly larger than the Earth, but in fact it is slightly smaller, having a diameter of 12,100 km (7,500 miles) just 656 km (406 miles) less than that of the Earth. It orbits the Sun at a mean distance of 108 million km (66 million miles) and in so doing, takes just 224·7 days to complete one orbit.

Venus comes closer to us than any other planet and is a really splendid celestial object, glowing like an enormous white lantern. At its brightest it has a magnitude of −4·0 which is brighter than any of the other planets. Records show that it is sufficiently powerful to cast shadows on moonless evenings. A small telescope reveals that the planet shows phases, just like those of the Moon. This is a characteristic of both Venus and Mercury, the other planet which lies between ourselves and the Sun.

## Mercury

Mercury itself is named after the winged messenger of the Greek gods. Its name is apt as it is an elusive little world which is always difficult to glimpse. For a start Mercury is rather small, having an equatorial diameter of only 4,870 km (3,020 miles). Furthermore it orbits the Sun at only 58 million km (36 million miles) and so never appears very far from it in the sky. It is these two factors which make it so very hard to locate, even in a telescope.

Town dwellers may have difficulty in spotting Mercury at all. One needs an horizon that is free of obstruction, and glowing street lighting will almost certainly render it invisible. However, given good conditions, presumably like those to which the Greeks

were used, the little world may be seen skirting a clear unobstructed horizon, and glowing with a faintly pinkish hue.

Like Venus, Mercury shows phases. When it is at position 1, it has its unlit hemisphere turned towards Earth and so is invisible. Mercury is then said to be 'new', and in this position is at *inferior conjunction*. As it moves along its orbit, more and more of the sunlight side becomes turned towards us, so it becomes

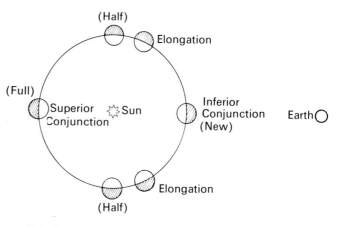

This diagram explains the phases of inferior planets, like Mercury and Venus. Such planets are difficult to see at all when at superior conjunction.

successively a crescent, a half (called *dichotomy*), three-quarters (*gibbous*) and eventually full, at position 3. When in the full position, known as *superior conjunction*, it is roughly behind the Sun and is very difficult to see, even with a good telescope. After superior conjunction Mercury passes through the same phases, but in reverse.

Now that Mercury's phases have been explained, it is easy to see why it is so difficult to spot. The planet is 'new' when at its closest to us. As it moves further away the phase increases and we see more of the

VENUS 1956 6" SPEC (VISUAL)

Venus showing different phases.

43

disc, but the apparent diameter decreases! The same happens with Venus, but being closer, larger and much more reflective, this makes little difference to its visibility. In addition Venus moves further from the Sun than its smaller neighbour, and so rises higher into the evening and morning skies before it is lost in the blaze of daylight.

Since both Venus and Mercury lie between us and the Sun, there are occasions when they can be seen crossing the solar disc. This does not happen every time either planet comes into the appropriate position in its orbit, since there are slight discrepancies in the angles of their respective orbits when compared to that of the Earth. Venus next will be *in transit* in the year 2004, while Mercury transits in 1986. The rarity of both events makes each very worthwhile watching.

All of the remaining planets lie beyond the Earth's orbit. Those known to the ancients are the next three in line: Mars, Jupiter and Saturn. Mars is easy to identify on account of its distinctive red colour and may sometimes become quite bright. Both Jupiter and Saturn have a slightly yellowish tinge.

It is most interesting to follow the slow progress of these outer planets against the background of the fixed stars. The plotting of their position, once every fortnight, will reveal some rather startling effects.

## Mars

The Greek astronomers were particularly interested in Mars as they watched its celestial wanderings. By plotting its course they noted that, from time to time, it started to reverse its motion, apparently going backwards in its path round the Sun. This backward or *retrograde* motion quite baffled them and so it is interesting to see why it occurs.

First of all, any planet which lies beyond the Earth cannot pass through inferior conjunction because it never passes between the Earth and the Sun. When Mars, for instance, is in the appropriate position (M1 in the drawing above right) and lined up with Earth and Sun, the Earth is in the middle which is a quite

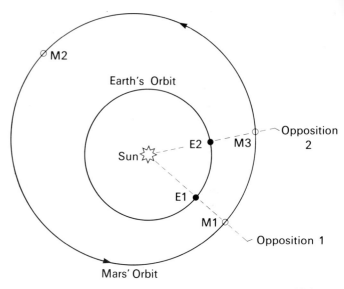

The orbit of Mars. Mars comes into opposition once every 780 days, whereupon we gain a favourable view of it. The Earth has to constantly 'catch up' Mars during its journey around the Sun.

different situation to that we hold with respect to either Mercury or Venus. When Mars is in this position we say it is at *opposition*. At opposition it is very favourably placed for observation since it lies on the opposite side of the sky to the Sun. For Mars, opposition takes place every 780 days, which is why there is a favourable observing period roughly every two years.

Mars comes into opposition each 780 days because both it and the Earth are moving, but at quite different velocities. The Earth takes only $365\frac{1}{4}$ days to complete one orbit, but Mars, lying at a mean distance from the Sun of 228 million km (140 million miles), takes much longer, about 687 days. This is Mars' *sidereal period*. In the diagram, let us start with the Earth (E1) and Mars (M1) at opposition. The Earth will have returned to its original position after one year (365 days). Mars, travelling more slowly, will have reached only point M2 at this stage, so that the Earth has to catch up Mars on its next revolution, which it does after 780 days (E2 and M3).

Now we can see why Mars appears to go back on itself as we plot its position in the sky. I have plotted the orbits of the two worlds and shown the

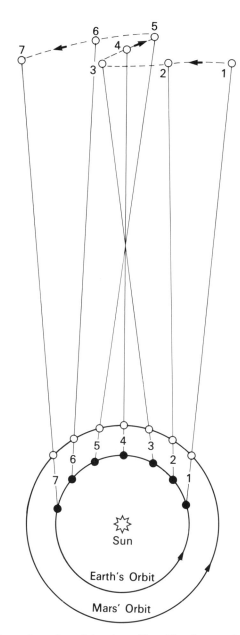

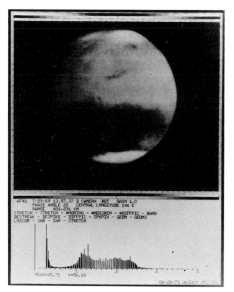

The surface markings of Mars as photographed from a distance of 401,231 km (249,313 miles) by Mariner 7 in July 1969.

The retrograde motion of the planet Mars. The diagram shows that between points 3 and 6, the faster-moving Earth passes Mars for a time, such that Mars appears to go backwards in its orbit for a time. The Red Planet spends 70 days in retrograde movement.

A closer view of Mars from 255,278 km (158,622 miles)

relative positions of Earth and Mars in our skies. Between points 1 and 3, Mars moves 'normally' across the sky; but between points 3 and 6 the faster-moving Earth is catching up with Mars and passing it so that for a time Mars retrogrades. Subsequently the normal movement will be observed. Out of the total orbital time, Mars spends just 70 days in retrograde movement.

Mars itself is quite a small world. It has an equatorial diameter of only 6800 km (4210 miles), which makes it only just over half the Earth's size. When viewed in binoculars or even a small telescope it is rather disappointing. Nevertheless its red colour can be well seen and it may be possible to spot the white polar 'caps' and the larger dark markings which are a characteristic of its surface; The next opposition occurs in February 1980, when the red planet will be in Leo, and any opportunity to look closely at the planet should not be missed.

## Jupiter

The huge planet, Jupiter was also known to the old skywatchers, although of course they were unaware of its great size. They, like us, knew it as a bright celestial object which shone with a yellowish hue, sometimes brighter than any of the fixed stars. They also knew that it, like Mars, did a backward loop in its orbit.

Because it is so far from the Sun—it has a mean distance of 778 million km (482 million miles)—it travels very slowly. This is in accord with one of Kepler's 'Laws of Planetary Motion' which states that the further a planet is from the Sun, the more slowly it moves. As a result the Earth has little trouble in 'catching up' Jupiter as it moves along, and we have oppositions of Jupiter at intervals of just over one year. As I write these words I can see Jupiter rising slowly in the eastern sky, against the constellation of Aries.

Jupiter like the other 'Outer Planets' is large but spins very rapidly on its axis. For this reason gravity has pulled out its shape into a very flattened spheroid, which is clearly visible in both a small telescope and in photographs. It takes Jupiter only 9 hours 55 minutes to spin once on its axis, although its equatorial diameter is eleven times that of the Earth (142,880 km—88,580 miles). It is so bulky that it is larger than all of the other planets rolled into one.

Jupiter's surface is covered in yellowish clouds which form a series of belts parallel to the equatorial plane. These clouds are quite reflective so that at its brightest it reaches a magnitude of −2·5; it is never difficult to spot. Galileo, in those early days of telescopic astronomy, noted that it was attended by four bright moons, which were named Io, Europa, Callisto and Ganymede. To watch the shifting patterns of these four satellites is a fascinating telescopic study. Not only do their relative positions change quite rapidly, but sometimes they are eclipsed by the great bulk of Jupiter, while at other times they cross the disc of the planet. There are another eight fainter satellites which cannot be seen with small telescopes.

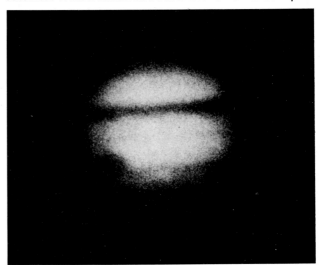

The cloud belts and Great Red Spot of Jupiter, photographed by Horace Dall. Note the polar flattening.

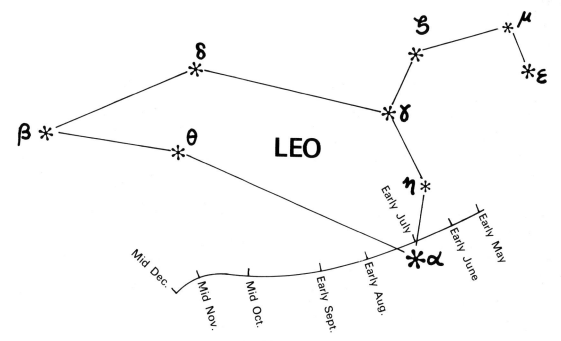

The path of Saturn during 1978.

In the star chart:
β θ δ LEO γ δ ℧ μ ε η α
Early May, Early June, Early July, Early Aug., Early Sept., Mid Oct., Mid Nov., Mid Dec.

## Saturn

The most distant planet known in ancient times was Saturn, planet of the rings. It lies some 1,427 million km (883 million miles) from the Sun at mean, and takes 29·5 years to complete just one orbit. At present it lies in the constellation of Cancer (the Crab) and is of magnitude +0·5, not as bright as Jupiter, but nonetheless easy to pick out. To help in finding the planet, I have drawn up a simple location chart showing its position for the next year. As can be seen, it moves very slowly against the stellar background.

Undoubtedly one of the most beautiful sights in the solar system is the view one gets of Saturn's rings in a telescope. Made up of tiny fragments of ice and dust, these completely encircle the body of the planet, a fact which was first recognised by Christian Huygens as long ago as 1655. The outer ring has an external diameter of 275,000 km (170,500 miles) and is separated by a marked gap from an inner ring of smaller dimensions. Huge though this ring system is, it is a mere 16 km (10 miles) in thickness.

Like Jupiter, Saturn has a number of satellites. In all there are ten. One of these, Titan, is the largest satellite in the solar system, having a diameter of about 4200 km (2,600 miles). It is easily visible in a small telescope and like the rings of Saturn was first seen by Christian Huygens in 1655. It is a very cold

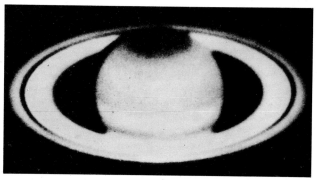

Saturn as photographed at Mount Palomar

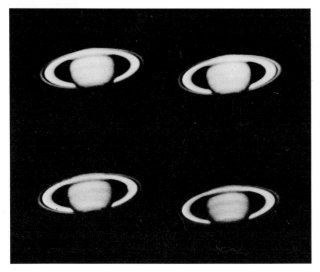

The rings and cloud belts of Saturn. The rings themselves are composed of tiny ice and dust particles.

world and has somehow retained an atmosphere, although not of a kind we could breathe.

One of Saturn's moons, Janus, was discovered only as recently as 1966. It is a tiny world, about 320 km (200 miles) across and actually lies between the inner ring and the body of the planet. If an astronaut could land (and survive) on this little moon, what a gallery seat he would have!

Saturn, like Jupiter, rotates rapidly on its axis, taking 10 hours 40 minutes to do so. Its axis, like that of Earth, is inclined to the plane of its orbit, in Saturn's case at about 26°. Since the rings circle the planet's equator, as it moves round the Sun we glimpse quite different views of the ring system. About every fourteen years the rings are 'edge on' at which time they are difficult to see at all, on account of their thinness. This last occurred in 1966 and will next happen in 1980. Between these dates we see the rings 'open' to varying degrees.

Saturn has one other claim to fame. It has the least

Different views of Saturn and its ring system as it moves with respect to the Earth. Sometimes the rings are wide open but at others are 'edge-on' and become almost invisible to us.

density of all the planets. It is less dense than water and so, theoretically, would float in the sea. Jupiter, too, has a very low density, as have the other giant outer planets I am going to mention shortly. This contrasts with the relatively dense, rocklike material from which the inner planets are made. Astronomers believe that the low-density outer worlds are composed very largely of ice and gas, particularly hydrogen compounds like methane and ammonia, which, because of the remoteness of the Sun, would be frozen.

## Uranus

As recently as the eighteenth century Saturn was

thought to mark the outer limit of the Sun's family. However, in 1781, Sir William Herschel, at that time an amateur astronomer living in Slough, Buckinghamshire, made a discovery which was to change not only the pattern of his life, but also of planetary astronomy.

While observing the stars of Gemini, he noticed a faint stellar object which he knew was not a star. At first he imagined it to be a comet, but a few simple calculations told him this could not be. It was, in fact, a 'new' planet, and was called *Georgium Sidus* by Herschel himself, in honour of King George II, who gave him a post of personal astronomer for his discovery. Subsequently, it was renamed Uranus.

Uranus is a frozen world with an equatorial diameter of 47,000 km (29,300 miles). It circles the Sun at the immense distance of 3,000 million km (1,860 million miles) and takes 84 years to complete one revolution. Since it was first found it has thus completed just over two orbits of the Sun. At the moment it lies on the boundary of the constellations Virgo and Libra, having a magnitude of about 6·0, rendering it on the limits of naked eye visibility.

In a small telescope it presents a greenish disc. This is crossed by faint dusky belts that are only visible in larger instruments. There are also five known moons, the largest having a diameter of about 1,000 km (620 miles). All need large telescopes to reveal them. One unusual feature of Uranus is that its axis lies almost in the plane of its orbit which means that Uranus would not have seasons of the sort we have. It must present its poles and equator alternately to the Sun, each pole experiencing a 42-year long 'summer' followed by a 'winter' of the same length. Even summer, for this planet, would be a cheerless season.

Herschel's discovery of a 'new' planet awoke the astronomical world from its slumbers. Soon people were avidly gazing at the heavens in case there should be more hidden worlds. One fact which emerged from a study of Uranus' orbit was that it seemed to have a slight 'wobble' that could really only be caused by the presence of another large world relatively nearby. In 1834, an amateur observer, Reverend I. J. Hussey, at that time rector of the parish of Hayes, in Kent, suggested it might very well be another planet. He wrote to the Astronomer Royal telling him as much, and got a rather rude reply for his pains.

Somewhat later, a young mathematics student at Cambridge, John Couch-Adams, did some calculations and came up with some data that suggested where this planet might be found. He also communicated with the Astronomer Royal (Sir George Airy), but was summarily ignored. Only when a more eminent French scientist came along with the same figures did Airy do anything. He started a search which proved fruitless. Then suddenly two French astronomers, Galle and d'Arrest, located the elusive world in almost exactly the spot predicted by Couch-Adams. So at last the eighth planet was tracked down. It turned out to be nearly a twin of Uranus, and was christened Neptune.

### Neptune

Neptune's orbit takes it about 4,500 million km (2,800 million miles) from the sun. Since being discovered it has not yet completed even one revolution: it takes 164 years to do so. Astronomers believe its surface temperature to be as low as −220 °C, so there can be little doubt that, like the other outer planets, it is largely composed of ice and frozen gases. It has two satellites, Triton and Nereid. The former was discovered by the British astronomer, William Lassell, another amateur observer, and is larger and brighter than any of the moons of Uranus. Nereid was found only as recently as 1948.

At the present time Neptune lies in Ophiuchus (the Serpent Bearer) and is of magnitude 7·7, so it cannot be seen with the unaided eye. However, binoculars will quite easily reveal its slightly bluish tinge.

In the early part of the present century, Percival

Lowell, a famous American astronomer to whom we shall again refer as we look in more detail at Mars, initiated a search for a planet even more distant than Neptune. This he did because Neptune, like Uranus, showed a distinct wobble in its orbit.

Lowell failed in his task, so in the 1920s another astronomer, E. C. Pickering, tried, but also failed to find anything. Then in 1929 two astronomers based at Lowell's observatory at Flagstaff, Arizona, took up the challenge. They took a large number of photographs of various parts of the sky and eventually, in 1930, spotted a faint starlike object which was moving against the star background. The discovery of this, the ninth planet, is attributed to Clyde Tombaugh of Flagstaff.

## Pluto

The new planet, christened Pluto, turned out to be far removed from what was expected. It is smaller than Mars, and measurements made with the only telescope then large enough to show it as a disc, the 500 cm (200 in.) reflector at Mount Palomar, suggest it to have a diameter of only 4,300 km (2,690 miles).

Pluto's 'year' is nearly 250 years long and its orbit is rather exceptional. It is decidedly eccentric and at perihelion it comes within the orbit of Neptune. It is unlikely that a collision could ever occur, however, as Pluto's orbit is also inclined to the plane of the other planetary paths. At present Pluto lies against the stars of Virgo and has a magnitude of 14. A large telescope is needed to find it.

At its furthest from the Sun, Pluto is about 6,000 million km (3,720 million miles), so it would appear that this marks the edge of the solar system. Or does it? A world the size of Pluto could not possibly cause a wobbling effect in the orbit of Neptune. It would be rather like a man trying to overturn an articulated lorry! So how do we explain this apparent anomaly?

One suggestion is that Pluto is very dense. Another is that it is larger than it seems, and that we are only

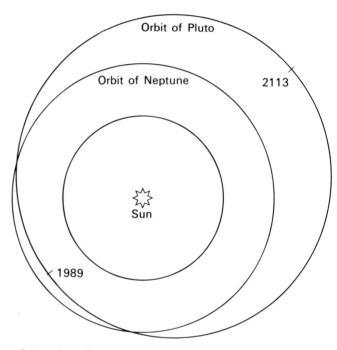

Orbits of the Outer Planets. Pluto's orbit takes it nearer the Sun than Neptune at certain times. Since its discovery, in 1930, the outermost known planet has not yet completed a single revolution of the Sun. This will not happen until the year 2178!

seeing (and measuring) the most reflective part of its surface. The third suggestion, and the one which is perhaps the most exciting, is that Pluto does not affect Neptune, but that some other, as yet undiscovered, planet does. In other words the finding of Pluto might have been quite incidental, the real cause of Neptune's odd behaviour being due to a large planet we have still to locate. Yet again astronomy provides us with a puzzle and with opportunities for finding the solution.

So we have come to the end of our planetary wanderings. This is not to say that there are no other members of the Sun's family. Meteors and comets are both part of the system, and there are also a large number of small rocky bodies, known as the asteroids, which circle the Sun between the orbits of Jupiter and Mars. Some of these can be picked up in quite a small telescope. All of these can be numbered among the celestial 'wanderers'. We shall be looking at these in a little more detail in chapter 8.

There have been new members introduced into the heavenly gathering, Man, in his wisdom, has constructed artificial satellites and space probes which have both circled the Earth, landed on the Moon and now even visited some of the outer planets. It would be a grave omission to neglect what these planetary visitors have found out about the planets. Much has been learned and many of the findings have been quite unexpected, as is so often the case with astronomical discoveries. Now seems the most fitting time to summarise the current state of our knowledge.

# SOME CLOSER VIEWS OF THE PLANETS

In March 1935, a manmade rocket lifted off and reached a speed of 1,120 km (690 miles) per hour. It fell back to the ground after reaching a height of 324 m (1,062 feet). This was a major success for one of the pioneers of rocketry, Robert Goddard, who had been persevering with his experiments since 1916. Since those early days rockets have become a part of our life, although it must be admitted that our experiences of them during the last war were not at all pleasant.

The first true steps into space were made by the Americans who, in developing their famous WAC-Corporal rocket at White Sands, New Mexico, eventually worked out the principle of a step-rocket, which involved using a heavy but very powerful rocket to set off lighter rockets, which then broke free of the cumbersome 'first-stage', blasting off on their own to greater heights.

It is one thing to get a rocket 'up', but quite another to get it into orbit or even away from the Earth altogether. Gravity constantly fights against the rocket in an effort to drag it back towards the Earth, and if the probe is to escape at all it must reach the critical velocity of 11 km (6·8 miles) per second, known as the Earth's *escape velocity*. Even to remain in orbit it has to attain 9 km (5·5 miles) per second. The earlier rockets were insufficiently powerful to reach such speeds.

Rather guarded comments continued to issue from

An old photograph of the rocket pioneer, Robert Goddard, with one of his successful rockets.

the USA during the early 1950s, but little was heard from the USSR. Then, out of the blue, the world learned that on 5 October 1957 the Space Age had arrived. Russian scientists had successfully launched a rocket that had placed a tiny artificial satellite, called Sputnik 1, into Earth orbit. It gave out a rather endearing 'bleep, bleep' signal which was broadcast by the BBC. Originally it took 96 minutes to complete one revolution and at its nearest point to Earth (*perigee*) it was only 218 km (128 miles) away.

I do not want to follow the course of rocket development in this book, as this has been done elsewhere (see the Bibliography). Suffice it to say that it took many years to perfect launching and docking operations to the stage when the first Apollo spacecraft landed Neil Armstrong and Buzz Aldrin on the lunar surface on 21 July 1969.

## Spacecraft to Venus

In 1973, the Americans launched a spacecraft towards our nearest planetary neighbour, Venus, on a trip planned to take it eventually past Mercury. This was not, however, the first trip which a rocket had taken in this direction.

The very first Venus craft was a Soviet probe. This set out in February 1961, but unfortunately mission control lost contact with it after it was only 7·5 million km (4·5 million miles) away. On 14 September 1964 an American craft called Mariner II flew past Venus, making certain measurements, but did not land on the planet's surface.

In March 1966, the Russian probe, Venus III, crash-landed on the surface, and a little later the Soviet scientists managed to land a craft on a descending parachute. This was a major step forward.

Venus is an odd world. It had hidden its surface from astronomers (and still does) by a cloak of thick clouds, and we had no idea of what lay beneath. Suggestions made included the idea that Venus' surface was covered by oceans, and also that it might be a world of choking desert. The Russian probe discovered it to be a very rough uneven world, very probably with craters like those of the Moon. A hint of this had come earlier by virtue of radar observations. American astronomers at the famous Jet Propulsion Laboratory had beamed onto Venus' surface with radar, and their analysis had strongly suggested a rough, mountainous terrain. This had now been confirmed ten years after their initial radar experiments.

The Russian probe which did land, measured the surface temperature as 260 °C. It also found that the atmospheric pressure was about twenty times that of the Earth. Venus is a particularly unpleasant place by any standards!

The hostility of our sister planet was further confirmed by the 1974 American craft mentioned above. This added to our knowledge of the atmosphere of Venus by detecting not only a great deal of choking carbon dioxide (which had been expected), but also droplets of unpleasant substances like sulphuric acid. The television cameras aboard the craft showed the layers of cloud very clearly and took the first pictures of the solid surface beneath. Hitherto there had been much uncertainty regarding Venus' rotation period, as it is perennially cloaked in its cloud mantle. Mariner established this to be 243 days. On the other hand, the upper cloud layers appear to rotate once every four days, giving Venus a very peculiar atmospheric structure indeed.

Mariner eventually left Venus and travelled towards the elusive Mercury, only this time Mercury could not escape its peering eye. What it revealed was quite unexpected. Here was a rugged landscape of great craters, deep valleys and enormous mountain ranges, just like that of the Moon. There were also vast plains, like the Caloris Basin, which are seared by the scorching rays of a very adjacent Sun.

Mariner 10 gradually moved away from Mercury

A close-up of the surface of Mercury taken during Mariner's second fly-by. The scarp shown is 298 km (185 miles) in length.

towards the Sun, and returned to its vicinity in September 1974, taking more photographs; it repeated the operation in spring 1975, when the instruments aboard were still in full working order. The little probe will presumably circle the Sun for ever, its useful life now ended.

### Spacecraft to Mars

Man has also turned his attentions to the other direction, towards Mars and the outer planets. The 'Red Planet', named after the god of war, clearly shows its true colour through binoculars. Our old friend Christian Huygens seems to have made the first drawing of the surface in 1659 and clearly depicted the dark, triangular-shaped feature we know as Syrtis Major. Soon after that discovery the polar caps were observed, which was why people started to ask whether there might be water on Mars, and possibly some form of life.

Many observers noticed that when the polar caps 'melted' during the summer season, a distinct darkening effect occurred in the regions adjacent to the shrinking cap areas. I have noted this myself during a number of oppositions, and it is almost as though

water released from the caps livens up vegetation that has been lying dormant during the winter.

Mars was known, quite early on, to have a tenuous atmosphere: clouds had been seen flitting across the

Christian Huygens' drawing of the planet Mars. The well-known dark marking, Syrtis Major, is shown.

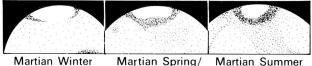

| Martian Winter | Martian Spring/Autumn | Martian Summer |

Seasonal variations in the polar region of Mars, as seen by the author in 1956. It seems that the summer 'warmth' releases moisture from the icy polar cap.

surface from time to time and photographs taken early in the present century clearly depicted these. The air appeared far too thin to support animal life, although there was a little oxygen present. However, vegetable matter might survive, so Mars has been considered a possible 'life base' ever since the eighteenth century.

Mars owes its vivid red colouration to the great expanse of desert which separate the darker regions. Good maps of Mars, such as that of E. M. Antoniadi (1902), clearly show these extensive tracts. Martian deserts are, however, far from torrid. A thermometer hung in the shade at midday would register a bare 2°C.

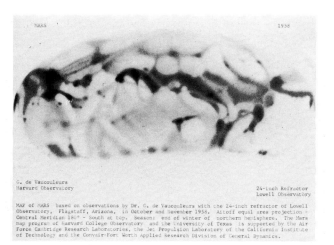

MAP of MARS based on observations by Dr. G. de Vaucouleurs with the 24-inch refractor of Lowell Observatory, Flagstaff, Arizona, in October and November 1958. Aitoff equal area projection - Central Meridian 180° - South at top. Season: end of winter of northern hemisphere. The Mars map program of Harvard College Observatory and the University of Texas is supported by the Air Force Cambridge Research Laboratories, the Jet Propulsion Laboratory of the California Institute of Technology and the Convair-Fort Worth Applied Research Division of General Dynamics.

Map of Mars drawn by Gerard de Vaucouleurs from observations undertaken at the famous Lowell Observatory, at Flagstaff, Arizona.

In 1877, an Italian astronomer, Schiaparelli by name, glibly announced that he had found dark, thread-like strips traversing the red desert regions — they were christened *canali*. Immediately the more imaginative workers began to conjure up pictures of intelligent Martians digging up the dry plains, making ditches to irrigate their crops. Things never looked back.

The task of trying to find out, once and for all, if there was life on Mars, fell to Percival Lowell of Flagstaff Observatory. He spent twenty years earnestly charting Mars' surface features, plotting hundreds of spidery canals on his maps. When he died in 1916 he firmly believed that Martians existed. Well, do they?

On 14 July 1965 an American spacecraft, Mariner 4, flew past the Red Planet and sent us back the first ever close-up pictures of that fascinating world. Surprisingly, it revealed a world scarred by enormous craters, some of them being clearly huge volcanic structures. Yet another surprise rocked the astronomer's world.

The clouds of Venus. Photographed by the Mariner 10 spacecraft in February 1974 from a distance of 720,000 km (450,000 miles).

Mariner also showed that the Martian atmosphere was even thinner than had hitherto been thought, which was a great disappointment to those who hoped to find life there. The air pressure at the surface is equivalent to that at a height of roughly 50,000 m (164,000 feet) above the Earth's surface.

In 1971, two Soviet vehicles wended their way towards Mars. Both landed a small capsule on the surface, but only one continued to function properly after

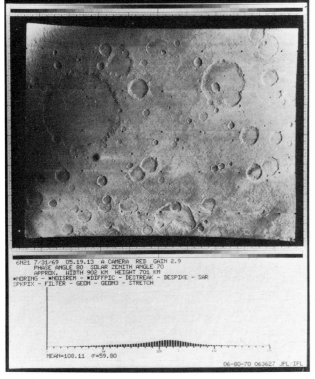

The crater-scarred surface of the planet Mars. A photograph taken by Mariner 7 at the end of July 1969. The width of the field of view is 902 km (560 miles).

landing. Nevertheless, it sent back valuable information regarding the conditions on the surface, including the wind speed, which is sometimes very high.

Later in the same year, a further American probe, Mariner 9, arrived and went into orbit round the planet. It sent back thousands of excellent photographs of the surface, including pictures of violent duststorms, and clouds which temporarily obscured the surface itself.

The most striking thing about the photographs was their lunar aspect. There are some enormous vol-

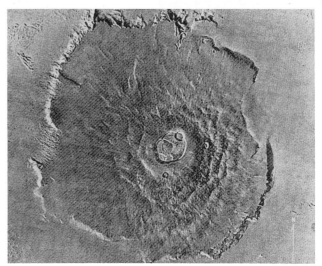

The enormous Martian volcanic centre of Olympus Mons, as photographed by the Mariner 9 spacecraft. The main 'cone' is about 550 km (341 miles) across and in places stands 4 km (2½ miles) above the surrounding plain.

canoes, such as the Nix Olympica, which is almost thirty times as large as the most expansive Hawaiian volcano on Earth. There are also deep canyons and sinuous channel features which really do closely resemble old watercourses, these latter being quite unlike anything seen on the Moon, suggesting that there has been moving water on the Martian surface.

The close-up photographs did nothing to explain the nature of the dark markings so clearly visible in a telescope, and drawn by Huygens in the early days of Mars exploration. Whether dark or light in hue, the majority of the terrain is cratered and there is certainly no sign of vegetable life.

In August 1976 the American space vehicle, Viking, actually landed on the Martian desert and scanned with its TV cameras, sending back a superb panorama of a rock-strewn desert terrain. The area it landed in, known as Chryse and shown on the Antoniadi map, is relatively flat. The probe reached it on the seventh anniversary of Man's first lunar landing, a fitting climax to the eleven-month journey. As I write this, American scientists are analysing the results of soil analysis undertaken by the probe's diggers, in an effort to find any evidence for past organic activity. So far their conclusions have been hesitantly negative.

View of the rocky terrain surrounding the Viking lander site on Mars. The photograph takes in the region of Utopia.

## Spacecraft to Jupiter and Saturn

Twenty years ago, if it were suggested that a man-made rocket would be visiting Jupiter or Saturn, the idea would have been laughed out of court. It just shows how science progresses beyond our wildest fantasies. In 1972 Pioneer 10 set out on its enormous journey towards the planet Jupiter. During its hazardous voyage, Pioneer had to negotiate the belt of asteroids which lies between Mars and Jupiter. Luckily it did this without mishap and it finally reached the vicinity of that distant planet in December 1973.

At its closest approach, Pioneer was a mere

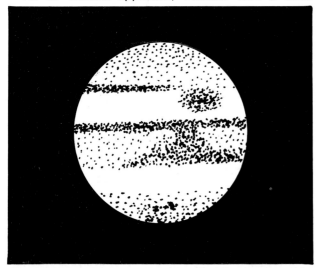

The cloud belts of Jupiter and the Great Red Spot as seen with a moderate telescope. The Spot is thought to be a vast meteorological disturbance.

130,000 km (80,600 miles) above Jupiter's surface. It sent back superb photographs of the cloud belts and discovered that the darker belts were related to downward-moving gas currents, while the brighter intervening regions sat above upward-moving gases. The Great Red Spot, an enormous pinkish feature that is clearly visible in even a small telescope, turned out to be some kind of vast meteorological disturbance.

Pioneer continued on its journey and has now passed further away from the Sun. It will finally escape from the solar system altogether. The Americans, with their usual foresight, enclosed a plaque, indicating where it was made, just in case some alien civilisation, millions of miles away in space, ever finds the little probe. It is our first celestial messenger, so 'Pioneer' is a very apt name.

Even before Pioneer 10 had completed its mission, another probe, Pioneer 11, had set off. This also reached Jupiter, this time in December 1974, took some more excellent photographs and also tried to take some shots of the four inner moons. It managed to secure a few, but they were all rather poorly defined. However, having completed its Jupiter mission, this probe is travelling towards Saturn, planet of the rings, and should reach it sometime in 1979. The astronomical world eagerly awaits its findings.

So much for man's wanderings among the planets. I want now to come nearer to home, to our own natural satellite, the Moon. Man has, of course, visited this world too, but before we talk about Apollo and Zond, we shall appraise the situation from a greater distance, from Earth. This was the starting point for all lunar exploration, one of the most fascinating of all branches of our science.

# OUR CLOSEST NEIGHBOUR

Since the dawn of mankind the silvery orb of the Moon has showered the night hours with its light. The protective glow was considered a divine gift by primitive peoples, who quite naturally felt particularly insecure in the dark, moonless nights, when marauding neighbours or prowling animals could attack them unawares. That the Moon showed phases and was not visible for part of each month has been known for thousands of years; but the reasons for its behaviour have been understood for fewer years than this.

Very early in history the monthly lunar cycle was used as a measure of time. It was also known that somehow the Moon regulated the tides. In an earlier chapter I explained why both of the inferior planets showed phases. The same explanation can be taken for our own satellite, the phases being strictly dependent upon the relative position of the Moon, Earth and Sun.

### Positions of the Moon

When the Moon is at position M1 it is 'new' and invisible to us, except that the unlit hemisphere may

The Moon and its phases. The relative positions of Earth, Sun and Moon define which phase we observe.

The Tycho region of the Moon.

be seen glowing faintly due to reflected 'earthlight', more often noticed when the Moon is a thin crescent. By the time it has reached point M2, it has completed one quarter of its monthly journey, and is thus said to be at 'first quarter'. At this time half of the disc is illuminated from our Earthly vantage point. Subsequently the Moon becomes 'gibbous' and eventually the whole of the Earthward-facing hemisphere is lit; we see 'full moon'. After full moon, the phases are reversed. We see 'last quarter' at point M4, until finally the Moon is new again.

One fact which confuses many people concerns the side of the Moon we continually see. It always shows the same hemisphere to us, although it spins on its axis, just as Earth does. How can this be?

The explanation lies in the fact that the axial rotation period is the same as what is called the *sidereal period*, the time it takes to complete one revolution round the Earth.

An old experiment illustrates exactly what happens. Stand someone in the centre of a room, move some distance away, and then walk slowly round in a circle, keeping your eyes focused securely on them. By the time you have arrived back at your starting point, you will have completed one revolution, but you will also have turned once on your 'axis'. However, the observer will not have had a chance to see your rear view at any time during the experiment.

The Moon's sidereal period is 27·3 days, as is also its rotation period. Yet our month is two days longer. Why is this? The answer lies in the fact that while the Moon is moving round its orbit, the Earth too has moved along. Consequently by the time the Moon has completed its circuit and arrived back at M1, the Earth has travelled onwards, and the alignment between Sun, Moon and Earth is not perfect. It takes just two days more for the Moon to reach the alignment point (M2) and it is then 'new' again. Thus the Moon's *synodic period*—the time between successive new moons—is two days longer than its

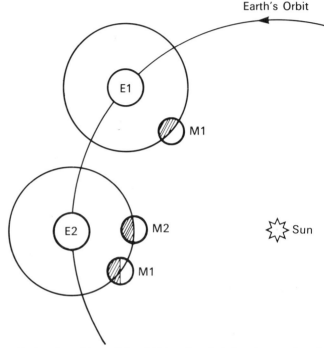

Explanation of 'synodic' and 'sidereal' period. Stand somewhere in the centre of a room a person or an object, walk slowly around this object keeping your eyes fixed on it until you have completed one revolution and come back to your starting point. By the time you have done this you will have 'rotated' once and completed one 'revolution' of the object.

sidereal period. It was the synodic period the ancients used to define one *month*.

What about the size of our satellite? In the early days some rather odd views were expressed. For instance, in the seventh century BC Anaximander held that: '. . . the Moon is a circle nineteen times as large as the Earth. It is like a chariot wheel, the rim of which is hollow and filled with fire . . .'

Larger than the mother planet it is certainly not. With a diameter of 3,476 km (2,170 miles), it is a mere quarter the size of the Earth. Lying as it does at a mean distance of 400,000 km (248,000 miles), it is by far the nearest natural object in our sky.

## Geography of the Moon

As we have discovered, Galileo and others saw the lunar landscape through their primitive telescopes. This view has not changed over the years and a first glimpse of the surface, through a modest telescope, usually leaves an impression of utter confusion. A great jumble of rugged mountains, ringed crater-like structures and grey plains greets the eyes. The shadows are black; the Moon is airless, grey and without halftones.

Initially the flatter, darker regions were believed to be oceans of water and were christened *maria* (seas). The bright, rugged, upland parts, known as *terrae* and covered in craters and mountains, inspired the early cartographers to award each a name, usually that of a famous scientist. Thus we have Copernicus, Kepler,

Julius Caesar and Tycho (Brahe) immortalised on the Moon.

During the eighteenth and nineteenth centuries lunar observers gradually constructed better and more accurate charts, many of which have become justly famous. In the early part of this century amateur observers became more and more numerous, and encouraged by such organisations as the British Astronomical Association—based in London—and the Association of Lunar and Planetary Observers—in America—the standard and volume of lunar observation increased several-fold.

Perhaps the epitome of painstaking amateur observation was Hugh Percy Wilkins, who toiled for nearly forty years at his telescope and drawing board, finally producing the magnificent *300-inch Map of the Moon*, showing a wealth of fine details, meticulously collected by observing through various telescopes, both large and small.

Apollo 16 panorama of the craters Aristarchus and Herodotus, showing the sinuous 'Schröter's Valley' and the plain of Sinus Roris beyond.

A telescopic view of the Goclenius region of the Moon, showing shadows cast by the walled structures known as craters.

Looking at the Moon is a fascinating experience, whether one intends to undertake serious observation or not. I never cease to be amazed at the architecture which exists on the surface. The cratered highlands are particularly striking, the bright rings standing out from the deep shadows of their interiors.

On the darker plains are gentler, volcanic structures, in particular the lunar 'domes'. These resemble 'shield' volcanoes, like those of the Hawaiian Islands, and many have small pits at their summits, like the Hawaiian examples. Then there are incised valleys, like the spectacular Herodotus Valley which cuts into the surface of Oceanus Procellarum close to the brilliant crater, Aristarchus. Winding ridges also traverse the maria and there are craters, mountain ridges and faults to be seen.

A description of what can be seen on the Moon would take a book to itself. It cannot be done here; in any case there are many fine books, well within reach of the average pocket—some useful ones have been listed in the bibliography.

Before man thought of sending rockets towards the Moon, he knew certain things about it. Firstly, the surface material was known to be dark, as the albedo was low. The maria reflect only about six per cent of the incident sunlight, while the brightest parts have an albedo of about double this. The way in which the surface reflected and polarised light, suggested that it might well be rather rough and possibly porous to a degree. Volcanic rocks seemed the most likely to occur. However, not all scientists agreed on this point: some firmly believed that a thick layer of rock dust shrouded the surface and that any spacecraft trying to make a landing would sink fatefully into it, as into a quicksand.

## Spacecraft to the Moon

The time eventually came when rocketry was sufficiently well advanced for there to be serious thoughts of sending a spacecraft Moonwards. Two questions really needed to be answered. Firstly, what did the other side of the Moon look like? Were there craters and maria similar to those we can see, and if so were there differences in their numbers or distribution? Secondly, there was the matter of settling what the Moon was made from, or at least what the lunar soil was like.

Failure dogged the first American attempts to reach the Moon in 1958. This was principally because the propellants used were not good enough to send the probes with sufficient force into the required orbits. The Russian team had better fuels and thus, in 1959, Lunik 1 raced on its way towards the Moon. It passed within 7,500 km (4,650 miles) of the surface and marked the first faltering step towards successful Moon travel.

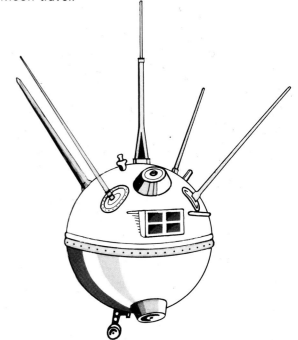

The Soviet Lunik One probe. This tiny spacecraft was less than 100 cm (39 in) in diameter but marked a most important forward step in space exploration when it was launched in 1959.

Lunik was only a tiny machine, less than 100 cm (50 in) in diameter, but it did send back to the Soviet space team important information concerning the lunar magnetic field. It showed that there was not one.

Lunik 1 now remains as a tiny member of the solar system. It was shortly followed by an American probe, Pioneer 4, which flew past the Moon at a distance of nearly 60,000 km (37,200 miles), and then in September 1959 the Russians replied by crashing Lunik 2 onto the lunar surface. It was tracked from Jodrell Bank (then the world's largest radio telescope and sited in Cheshire, England) all the way until eventually the signals gave out at 21 hours, 02 minutes and 23 seconds on 13 September. It had impacted on the lunar ground and destroyed itself.

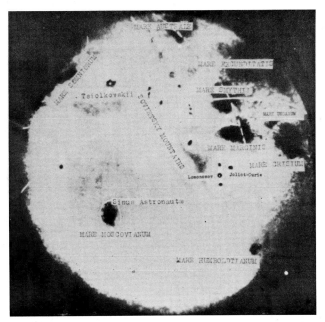

One of the first photographs of the Moon's farside sent back by the Soviet probe Luna 3, in 1959.

To cut a long story short, several more successful craft were launched in the ensuing years. One, in particular, was of great importance. This was Lunik 3, which went round the back of the Moon and sent to Earth the first photographs of the Moon's 'behind'. They were first screened on the BBC TV astronomy programme 'Sky at Night' on 26 October, and I can clearly remember Patrick Moore's almost uncontrollable enthusiasm for this wonderful view.

Lunik showed that there were indeed craters on the averted hemisphere. In fact, there were more highland areas than there were on 'our' side and fewer maria. For the first time many of the features near to the Moon's edge or 'limb', were revealed in their true undistorted guise.

Quite a gap separated this successful flight from the next Russian effort. It was not until 1965 that the Soviet craft, Zond 3, sent back greatly improved shots of the hidden side. These allowed Russian cartographers to draw up the first-ever accurate chart of the Moon's other side; a task which had taken 350 years since man first saw the lunar landscape through a telescope.

Meanwhile the American space programme had been striding ahead more successfully, and although the early Ranger rockets failed, subsequent models met with great success. In July 1964 Ranger 7 reached the Moon, its TV cameras were switched on by computer signal and space scientists received the first real close-up shots of a part of the surface near to the crater, Guericke. The first frames started at an altitude of 2,400 km (1,500 miles), and the last frames were of tiny pits in the lunar soil, just before the Ranger hit the ground.

One fact which emerged from these photos was that the surface near Guericke crater was not shrouded in a thick layer of dust. Very clear-cut crater structures were visible right up until the point when the last frames were obtained, only 0.2 seconds prior to impact. So one question had been answered.

More superb pictures were shown by later Rangers, one of which crash-landed inside the large crater, Alphonsus. In January 1966 the Russian probe Luna 9 actually soft-landed on the wide plain, Oceanus Procellarum, and within minutes of touchdown returned the first-ever 'live' photos from the Moon's surface. Again, it was clear to see that there was no layer of dust; the terrain was flattish and rocky and looked remarkably like some of the Earth's volcanic regions. A little later, an American craft, Surveyor 1, landed in the same general region, and sent back 10,000 shots of very superior quality. Man had most definitely conquered the space between Earth and its nearest neighbour!

The next step, as seen by the American scientists, was to send *men* to our satellite. Sending a rocket is one thing. Landing it and getting it safely back to Earth is another. This required a far higher degree of skill altogether.

Nevertheless, scientists always seek to progress and they persevered with their rocket technology: the Americans with a view to landing a man in their Apollo spacecraft, and the Russians, intending to perfect the art of soft-landing delicate instruments, but unencumbered by human cargo.

### The First Landing on the Moon

Who can forget the 21 July 1969? This, the day when two astronauts actually set foot on the lunar soil, is indelibly imprinted on my memory, and I suspect on that of hundreds of other astronomers, both amateur and professional. Apollo 11 had successfully landed on Mare Tranquillitatis and shortly afterwards, as we watched our TV screens, Neil Armstrong and Edwin 'Buzz' Aldrin slowly backed down the lunar module's stepladder, finally taking their first tentative steps, at 2 hours, 56 minutes GMT on 21 July 1969. An historic moment indeed!

The first thing Armstrong did was to collect the 'contingency sample', a small specimen of lunar soil which could be quickly taken back to the module if something went drastically wrong, requiring the two men to make a quick getaway. However, no 'little green men' came into view, and neither did the module disappear in a desert of dust. Everything went according to plan.

Both astronauts walked quite happily about in the low gravity conditions of the Moon—the escape velocity is a mere 2·4 km per second (1·5 miles per second)—and their bootprints were clearly seen in the photographs they took and those which came back via television. So, the surface was hard and the dust layer very thin, a few cm only. The rocks they collected were eventually returned to Earth on 24 July, when they splashed down in the Pacific Ocean to a tumultuous welcome.

A dream had now been fulfilled. Man had been to the Moon and had brought back, not only his impressions of the lunar environment (described by Aldrin as '... magnificent desolation ...') but also samples of the surface rocks. These were eventually dealt out to various scientific institutions and laboratories, and found to be similar to terrestrial volcanic rocks, such as basalt. However, there were certain differences which marked them quite distinctly as lunar. In particular there was the absence of any water combined with the minerals, something which is normal for terrestrial rocks. This fits in, of course, with the Moon having no atmosphere of its own, due to the very low surface gravity.

### Space Programmes to the Moon after 1969

The rest of the story has been reported in the Press and a large number of books have appeared dealing with man's adventures up there in space. More Apollos followed the first one to land. One nearly met with disaster and had to return to Earth without landing on the Moon at all. The rest, however, were tremendously successful. In the later ones, a lunar 'buggy', a motorised lunar Land Rover carried the two astronauts as far as the crashed hull of one of the earlier Surveyor probes: a distance of many km. More rocks were collected and were analysed back on Earth. Experiments were set up on the surface, particularly dealing with the recording of seismic waves, or 'moonquakes'. Then, of course, it must be remembered that the command module—the craft from which, and back to which, the lunar module was launched during each trip—was continuously circling the Moon while the astronauts were on the surface, and taking thousands of superb high-resolution photographs. These have been used to draw up the finest lunar charts ever seen by Man.

While all this was happening, the Soviet space programme had been continuing, if somewhat less spectacularly. Luna 16 arrived on Mare Foecunditatis in 1970 and managed to drill into the surface to a depth of 35 cm (14 in), recovering a core of the subsurface material. This was later brought back to Earth and analysed. Chemically, it was very similar to many of the American samples and helped to confirm that lunar surface rock is very like basalt, the commonest volcanic rock on Earth.

Later Luna 17 landed, this time on Mare Imbrium. The craft contained sampling equipment, a laser-reflector and highly sophisticated radio equipment. Three months after landing, in other words by March 1971, the transporter had surveyed a strip of lunar country over 3 km (1·8 miles) long by 65 m (213 feet) wide. The mission was an undoubted success, and showed that the Russian effort to perfect unmanned exploration of other worlds was paying high dividends.

Today the Apollo programme has been completed. The next stage in the American quest is the perfection of the 'space shuttle', a rocket-powered craft which can travel to and from an orbiting space 'station'

above the Earth. This would give tremendous help towards easing the problems of rocket launching from Earth, where the gravity is high, and will undoubtedly be a major advance in space technology, if it works.

What have we learned about the Moon? Well, we know that there is no deep dust layer. The surface rocks are volcanic and are covered in places by fragmental matter, including glassy particles unlike anything of terrestrial origin. There are frequent moonquakes, many apparently being triggered off by lunar landslides set up near the walls of craters, where the loose debris is rather unstable. There is no atmosphere, no magnetic field and there appears never to have been any water.

It is not a friendly place, by any standard. If Man wants to set up laboratories or observatories there, he will have to take all his air and food with him. The Moon 'base' must surely still be a long way off, so the science fiction writers still have time to pursue their fantasies for a while.

Scientists are not a hundred per cent sure of how the craters were formed. Personally, I favour their having a volcanic origin, at least the larger ones. However, there can now be little doubt that many of the smaller pits were formed by the impact of meteorites, small fragments of rock which constantly bombarded the airless surface of the Moon. Like so many aspects of astronomy, although we have found out much more about this one part of the science, we still do not have all the answers. I believe I am on pretty firm ground in saying that it is unlikely that we shall for many decades to come. The Moon is still well worth watching, and may still have surprises in store. In any case, there are few more fascinating things than a view of the brilliant Moon through a telescope.

# OTHER LIGHTS IN THE SKY

## Comets

In 1973 astronomers forecast that a brilliant celestial visitor would grace our skies during late December. Disappointingly this vagrant, called Comet Kohoutek, was much fainter than had been predicted, and never became a bright object. With not a little embarrassment, the same astronomers had to explain away the comet's behaviour to the Press.

While comets do sometimes disappoint, on other occasions they surprise and delight the eye. Thus in 1976, West's Comet appeared, becoming a prominent object in the sky for some months, and was by far the brightest comet seen for several years.

Some comets look rather like the notorious UFOs which were so prominently reported in the late 1950s and early 1960s, appearing simply as bright balls of diffuse light. Others, however, are graced with fine illuminated 'tails' that stretch across the celestial vault. The last really spectacular comet appeared as long ago as 1910, so we are surely long overdue for a brilliant visitor.

What, then, is a comet? Is it a part of the Sun's family or a casual wanderer from further afield?

First of all, comets are made from extremely flimsy material: tiny particles of rock and ice held together in an envelope of tenuous gas. A classic comet has a bright *nucleus* surrounded by a *coma*, from which extends a glowing *tail*. The latter may extend for as much as a third of the way across the sky.

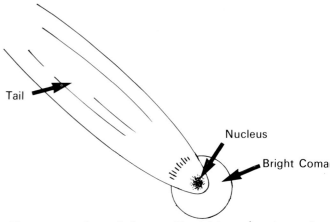

Tail

Nucleus

Bright Coma

The structure of a typical comet. The bright nucleus is usually surrounded by a fainter 'coma' with the tail streaming out from the nucleus.

Most comets develop a tail only when they approach the Sun, that is, at perihelion. It seems almost as though the gases which make up the cometary tail are extruded from the nucleus in response to the heating-up process it experiences while near the Sun. Accordingly many cometary tails change their form after perihelion passage, and some disappear altogether.

The tail itself always points away from the Sun. Thus during its journey towards perihelion, the tail follows the nucleus in what we might call the 'normal' manner. After perihelion, however, when the comet has swished round the Sun at great velocity, it moves tail first. Originally it was thought that the pressure exerted by light was responsible for this curious behaviour, but more recent research suggests that magnetic and other particles sent out by the Sun affect the cometary material in this strange way. The unpredictability of comets is one of their main attractions.

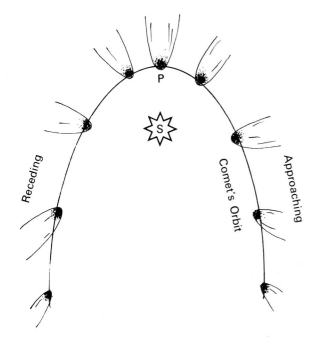

During its path round the Sun the cometary tail always points away from the mother star. Thus after perihelion, the comet appears to move 'tail first.'

A large number of comets visit us on their way through space. Some return again, others do not. Those which do, the periodic comets, are true members of the solar system and have elliptical orbits like those of the planets, but generally of much greater eccentricity. The orbital size and shape varies enor-

mously and so, naturally, does the length of time which must elapse before each periodic comet returns to the neighbourhood of Earth.

Non-periodic comets have parabolic or hyperbolic orbits. Such bodies are true space wanderers for they grace our skies once, never to return. The disappointing Comet Kohoutek was of this type.

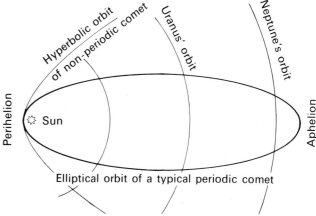

Cometary orbits vary from nearly circular, through very eccentric elliptical to hyperbolic.

### Halley's Comet

The most celebrated of all is undoubtedly Halley's Comet, so named after the astronomer who realised it was a true periodic visitor and predicted one of its returns. The modern history of the comet begins in 1682 when the German astronomer, Dorffel, and astronomers at the Royal Greenwich Observatory, then situated just outside London, recorded a comet during August. Among the British-based observers was Edmond Halley, subsequently to become Astronomer Royal.

In the summer of 1682, the comet was very striking and had a brilliant tail. At the time it was not appreciated that comets were true members of the Sun's family. However, Halley, being preoccupied with the comet, undertook some calculations and eventually wrote an important paper which he presented to the Royal Society. His findings were that the 1682 comet was the same as one which had been recorded both in 1531 and 1607. He predicted its return for the year 1758.

Halley did not live to see whether or not his prediction came true. Needless to say, it did. On Christmas Day 1758, an amateur astronomer living in Germany, made the first sighting of the comet as it approached the Earth. It reached perihelion in March 1759 and by that time had been recorded by astronomers scattered all over the world. It was unanimously agreed that it should be named Halley's Comet, in honour of its predictor.

Halley had died in 1742, but his comet reappeared in both 1835 and 1910. Its orbit reaches out beyond the path of Neptune and at present it is approaching the Earth. It is next due in our part of the solar system during 1986, so many readers may be privileged to see the great comet in all its glory.

In 1910, the year when Halley's Comet last appeared, an even more spectacular object graced the skies. Another comet passed close to Earth which was so brilliant that it could be clearly seen in day time; consequently it was called the 'Daylight Comet'. As is so often the case with very brilliant comets, it was a non-periodic object, and will never be seen again. It has long since passed out of the solar system, along a parabolic path.

In olden times the appearance of a comet produced almost instant panic and many very strange beliefs surrounded these celestial visitors. The Romans, for instance, believed that the comet of 488 BC was Caesar's soul being carried to the gods. According to Siculus, another comet in 372 BC (which was seen and described by the great Aristotle) announced the fall of Sparta. Comets also appeared at the times of the deaths of Constantine the Great (AD 336), Louis II (AD 875) and Richard II of England (1198). At least two popes returned to their maker while comets were visible in the sky! So per-

haps there is some justification for this primitive fear. Will something terrible happen when Halley's Comet returns in 1986? We shall just have to wait and see ...!

## Faint Comets

While brilliant comets are relatively infrequent, fainter ones are quite common. Many are too faint to be seen except with quite large telescopes, but every year astronomers, both amateur and professional follow the paths of these periodic visitors and the changes in form their bodies undergo. Over one hundred short-period comets are known. Many of these have paths which reach the region of Jupiter's orbit at aphelion. The giant planet evidently plays an important part in their life history and while it is unlikely that Jupiter itself gave birth to the comets, its strong gravitational attraction must have affected their orbits.

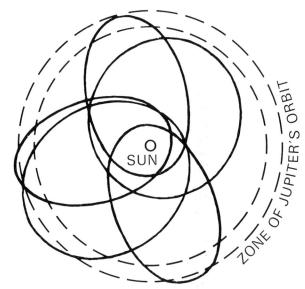

Jupiter's family of comets. The gravitational attraction of the largest planet clearly has had a profound effect on some passing comets and has provided the giant planet with its own family.

One of the most reliable and long-lived of the faint comets is Encke's Comet. This well-known object has a period of 3·3 years and has been observed regularly for over 130 years. Like most comets of this kind, its magnitude is low, and at present it is of the twentieth magnitude; thus it can be seen only with a very large telescope.

One or two much more spectacular objects have streamed across the sky in recent years. In November 1956 two Belgian astronomers recorded a comet which, by April 1957, was to become a prominent object in our skies. It was a particularly beautiful sight through binoculars, and during the latter part of its visitation it developed an odd 'reversed' tail that became known familiarly as its 'beard'. This comet, named Arend-Roland, remained visible until nearly the end of April. In the same year another one, Comet Mrkos, was a bright object during August. So it was a good year for comet observers by any standards.

Since 1957 the only really inspiring comets have been Bennet's, which appeared in 1970 and was particularly prominent in April of that year, and West's Comet which I have already mentioned. The former's tail stretched for 20° across the celestial vault and its bright nucleus rivalled the brightest star of Cassiopeia.

### Meteors and Meteorites

There are other kinds of celestial visitor, of which the meteors are the most spectacular. Very often, in fact, these are confused with comets. However, whereas the latter are periodic and longlasting a meteor appears as a shortlived flash across the sky, like a giant firework.

Meteors are simply small fragments of rock which wander near enough to the Earth to be caught in its gravity field. They are pulled downwards through the atmosphere, whereupon they burn up due to friction, glowing as they do so. Most are so small that they never reach the surface, but sometimes unusually large rocks come along, and these may crash onto the Earth. We call these objects *meteorites.*

Large meteorites are rare. One of the largest of recent times fell in a remote part of Siberia during 1908, flattening forest trees over an area of some 80 km (50 miles) in diameter. A large impact crater was formed and pieces of the object were collected for study. Much more recently (on 8 March 1976) a brilliant 'fireball' was seen. This exploded and showered fragments over a wide area of north-eastern China. The largest fragment collected weighed 1,770 kg (3,900 lb).

The most spectacular recent fall in Britain occurred on Christmas Eve 1965. On that evening there were many reports of a brilliant 'fireball' moving across the sky, a particularly large number of sightings being made in the English Midlands. Fragments of this object were eventually found in and around the village of Barwell, in Leicestershire, and a reconstruction

Comet Arend-Roland, as photographed in 1957 by E. Lindsay. Note the peculiar 'beard' or forward tail of this comet.

from the fragments listed suggested that the original meteorite weighed about 90 kg (200 lb).

One rather amusing story, recounted to me by a relative of the man involved, surrounds what may well have been one of the largest pieces of the Barwell meteorite. The said gentleman, on coming out to start his car on Christmas morning, found a large hole in the bonnet and a correspondingly guilty looking stone in the engine. Believing vandals to be responsible for the attack, he blasphemed vehemently and hurled the offending object into the surrounding undergrowth. Shortly afterwards the story came out and this Barwell man realised the nature of that 'stone'. It was never located.

Meteorites themselves are of two main kinds: stony ones known as *aerolites* and iron—nickel objects called irons or *siderites*. Many museums have chunks of both kinds, but it takes an expert to decide whether any suspicious-looking object has an extraterrestrial origin. Geologists believe that most meteorites hail from the 'asteroid zone' which is a zone of rocky bodies situated between the orbits of Mars and Jupiter. In this zone there are many large chunks of rock, known as the asteroids, or 'minor planets', which orbit the Sun just like the major planets, and which almost certainly represent the remains of an old planet which somehow disintegrated. The tell tale proximity of the asteroid zone to the giant planet Jupiter points to the strong gravitational pull of that planet as having been the cause of the catastrophe which must have befallen the original planet. Meteors may well represent the smaller sand-sized fragments left after the event.

## Biela's Comet and its Meteorite Connection

Since comets and meteors are both composed of small fragments of rocky matter, it would be surprising if there was no connection between them. Indeed, there is one piece of evidence which points to a very close connection between a comet and some meteors. This link is illustrated by the story of Biela's

Comet, whose first recorded appearance dates back to 1772.

Biela himself recorded this comet in 1826. Calculations he made after its movements had been closely studied suggested it to be a short-period comet with a period of 6¾ years. There was also evidence to suggest that it was the same comet as one which had been seen in both 1772 and 1806.

The comet duly reappeared in 1832, was missed in 1839, but was again recorded in 1845. However, during the latter return it surprised astronomers by

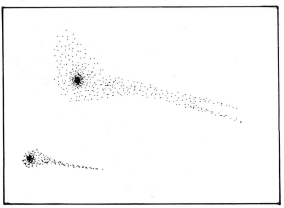

Biela's Comet. This unusual comet actually split into two parts, at one stage the two main parts being linked by a 'bridge' of tenuous material. Eventually it broke up and the periodic shower of meteors, the Bielids, is thought to be all that remains of it.

splitting into two parts, which were sometimes joined by a tenuous 'bridge' of material. After perihelion the comet duly faded and its next return was eagerly awaited. In 1852 it was seen again, but this time the two parts were separated by an even greater distance than before. In 1859 Biela's Comet was not seen at all, possibly due to adverse observing conditions, nor was it seen in 1865 either, although it should have been an easy object to observe.

Astronomers who had searched in vain during 1865 did so again during the next predicted return in 1872. In that year, however, there was a very rich display of meteors which could have been purely coincidental, had not the same shower subsequently appeared each time that the Earth crossed the path of the 'dead' comet. The inference is obvious; the old Biela's Comet had disintegrated into meteor-sized particles. The shower next returned in November 1977.

Meteors very frequently occur as showers. Large numbers of them streak across the sky during a short period. A rich display is a very fine sight and over a hundred meteors may be counted in an hour's observing time. In general each meteor appears to have its origin at a definite place in the sky, known as its *radiant point*. Thus the famous 'Perseid' shower has its radiant in Perseus, the 'Leonids' radiate from Leo and so on. Of the annual meteor showers the most spectacular are the Quadrantids (which radiate from near Ursa Major, between 1–6 January), the Lyrids (19–24 April and 10–24 June), the Perseids (25 July–18 August), the Orionids (16–26 October), and the Geminids (7–25 December). The main hindrances to watching and counting meteors, weather aside, is the presence of the Moon, which easily drowns the trails in its glow. When conditions are favourable, however, many fascinating hours can be spent counting the numbers of meteors and plotting the paths of meteors on a star chart, in an effort to define the radiant point.

## Zodiacal Lights and Aurorae

Not all the celestial 'lights' are as well defined as comets and meteors. Others take the form of glows in the sky. The *zodiacal light* comes into this category. It may sometimes be seen soon *after* sunset in March, or just *prior* to sunrise in September. It takes the form of a faint glow spreading up from the hidden Sun towards the zenith. This odd phenomenon is believed to be produced by thinly spread particles sent out from the Sun which are somehow illuminated by the solar light.

Much more spectacular is an *aurora*. Unfortunately one has to live rather far north or south ever to see a good display. When I lived for a few years in Elgin, Scotland, I was fortunate to see several very fine displays. Each started from the horizon with a gradual brightening of the sky, eventually reaching almost to the zenith. As the aurora developed, greenish 'tentacles' of light would surge even further across the sky, pushing outwards and then retracting again like silent ghostly fingers. At the height of the display, shimmering curtains of greenish light hung from the

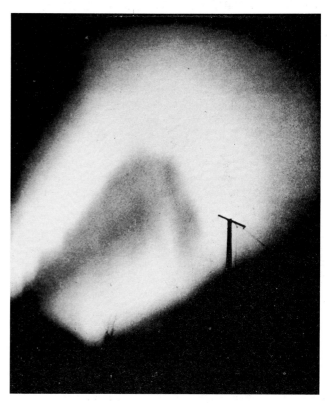

A bright auroral display as seen from Alaska

An auroral display as seen from Scotland.

The effect is incredibly awesome and is something which once seen is unlikely to be forgotten. Its origin is found in the excitation of magnetic particles located high in the Earth's atmosphere. The glow is produced by radiations sent out from the Sun. Since magnetic activity is very strong at the poles, auroral displays are best seen in high latitudes.

Most people will have heard of 'Flying Saucers', otherwise termed UFOs (Unidentified Flying Objects). Some of these strange objects may be mis-reported meteors, others can be variously put down to natural illumination, aeroplanes, meteorological balloons, artificial satellites, high-level ice clouds, or some even to pure imagination. A very small minority are difficult to explain at all. It is these about which some people like to fantasise.

Do they perhaps come from other planets or other star systems? There is no room in this book to argue the point, but what we now can do is to see what does lie beyond the solar system and even beyond our own star system, in the depths of the universe.

sky, their pattern of 'folds' changing from minute to minute, as if they were being disturbed by an unfelt celestial breeze. Gradually the spectacle would fade over a period of an hour or so, and eventually disappear as quietly as it had come.

# THE GALAXY

## The Milky Way

Everyone who has looked into a clear star-studded sky must have seen the Milky Way. If asked to describe it he or she would find it hard to better the description of Ptolemy who, writing two thousand years ago, described it as '... a zone which is almost everywhere as white as milk ...'. Galileo, who studied it with his tiny refractor, found that it was composed of thousand upon thousand of individual stars.

If we trace its path across the heavens, first from Cassiopeia, we find that it wends its way through Perseus, Auriga, Gemini and then down through the heavenly Dogs, Canes Minor and Major, before moving into the southern constellations of Argo, Crux Australis and Centaurus. Moving back through more northerly groups it crosses Scorpio, Sagittarius, Aquila and Cygnus before reaching Cassiopeia again. The most brilliant section is found in Sagittarius. This is a pity, for it means that northern observers never get a really good view, since this constellation can be seen only low down during summer evenings when the sky is never completely dark.

The Milky Way is a region where the stars are more closely packed than elsewhere. All the stars within it constitute what is called a *galaxy*. About 170 years after Galileo viewed the Milky Way, Sir William Herschel, a talented musician who became avidly interested in astronomy, and who fortunately could afford to support his astronomical work via his musical activities, started to study the Milky Way with telescopes which he had himself designed and built. These were far superior to any previously existing, and although compared to modern instruments they were of relatively small aperture, he was able to make several important forward strides in our understanding of astronomy. While Herschel had no means of measuring the distances of the stars themselves, by studying their numbers in some detail he concluded that the Milky Way had the shape of a telescope lens and he concluded that the Sun must lie near to its centre. He had no proof for the latter suggestion and was simply following the consensus view of his predecessors.

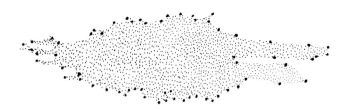

The form of the galaxy according to Herschel. He concluded that the Sun must lie near to the centre.

The idea of a centrally placed Sun is, of course, very reasonable. However, the density of the stars along different parts of the disc is obviously variable, the Sagittarius region being particularly rich in stars. This could either be sheer coincidence or their could be some other explanation. This doubt led a Harvard astronomer, Harlow Shapley, to enquire more closely into the subject.

## Globular Clusters

Working at Harvard during the early 1920s, Shapley made a study of what are known as *globular clusters*. Earlier in the book we looked at the Pleiades and the Hyades, *open clusters* of stars which form local groups scattered about the universe. The two I mentioned are easy to see with the unaided eye and are quite spectacular objects in binoculars. The globular clusters are quite different. These globular groups are large star families, in which the stellar concentration is much more dense. Only a small number of such clusters can be seen at all without telescopic aid.

The brightest of these lie south of the equator. Finest of all are Omega Centauri and 47 Tucanae which lie well south of the visual limit of observers

The great globular cluster M13 Herculis. This cluster of stars can be seen from the northern hemisphere as a faint hazy patch in the constellation of Hercules.

in the northern hemisphere. There is, however, one globular which can be seen from northern latitudes. This lies in the rather faint constellation of Hercules, and is known as M13 Herculis (see bottom left). It is not an easy object to find as it is rather faint and the stars of Hercules itself are a trifle inconspicuous. However, it is just visible without optical aid, lying between Zeta and Eta Herculis and rather nearer to Eta. A small telescope reveals it as a dim fuzzy patch, but with larger instruments its stellar nature is finely shown.

Globular clusters are not particularly numerous. Just over a hundred are known and maybe double that number actually exist. What puzzled Shapley was that the clusters were nearly all found in southern latitudes. This could surely be no mere chance? There must be some rational explanation.

Measuring the distance of these remote objects was a task of some difficulty. However, it was found by Shapley that within many of the globulars there were variable stars of a type known as *RR Lyrae stars*. These are regular variables with short periods, like the Cepheids, and they all appear to have roughly the same luminosity, namely ninety times that of our own Sun. So once the magnitudes of any RR Lyrids within the globulars could be established, it would not be a difficult calculation to measure their distances.

Shapley made this calculation for many of the globulars which contained RR Lyrae stars and he found firstly that they were very distant, and secondly that they had a very lopsided distribution indeed. The globular cluster in Hercules turned out to be 34,000 light years away and even the much brighter cluster, Omega Centauri, lay at the immense distance of 22,000 light years. It is this remoteness which renders them such faint objects in the sky. In fact many of them shine with the light of a million suns and on average they contain about 100,000 individual stars.

The fact that the majority of the globulars lie in the southern constellations of Scorpio and Sagittarius led Shapley to realise that the Sun could not be centrally

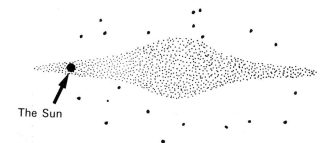

The actual shape of the galaxy showing the position of the Sun, well removed from the central regions. It is this eccentric position which gives us our view of the Milky Way.

The Sun

situated within the galaxy. Our asymmetrical view is a result of our situation well away from the galactic nucleus. It was also possible, for the first time, to assess the dimensions of the system. It was found to have a diameter of 100,000 light years and to have a bulging central nucleus with a 'thickness' of about 20,000 light years.

## Nebulae

Oddly enough another important step which had been made well before the time of the famous Harvard astronomer was taken by a French observer of not stars but comets. Charles Messier lived during the eighteenth century and he published an important catalogue of celestial objects listing over one hundred items. The great globular cluster in Hercules is, for instance, listed as item number 13, and hence is called M13, the 'M' relating to Messier's catalogue.

During his sky sweeps in search of comets, Messier was constantly being held up, as he found large numbers of diffuse objects which at first sight appeared to be 'new' comets. However, after studying them for a few days, he quickly realised that they showed no movements, as did typical comets, and thus could not be cometary by nature. So that he might save himself the frustration time and again, he set about compiling a list of these objects. He called them *nebulae*, after the Latin word for a cloud, and in 1781 he published his list. Number M1 is an object lying in Taurus, while the hazy patch in Andromeda

which we have already noted in an earlier chapter, is Messier's object number 31. His catalogue numbers are still used today.

When Herschel started studying the Milky Way he also became interested in nebulae. Some, he found, did seem to be clouds of shining gas, without any signs of stars within. However, others certainly did contain stars in great numbers. In particular the Andromeda 'nebula' (M31) was partly stellar and there were others besides. The most typical of the 'gassy' nebulae appeared to be that in Orion's sword, object M42 in Messier's catalogue, known as the Great Nebula in Orion. Herschel made the bold guess that while the gaseous clouds were within our own galaxy, the starry ones were well beyond it.

At this time photography was still an invention for the future. Astronomers had to rely on visual observations for their researches. During the middle part of the nineteenth century, Lord Rosse, of Birr Castle in Ireland, constructed and erected a huge reflecting telescope with a mirror nearly 125 cm (50 in) in diameter. He did this with the specific intention of studying the gaseous and starry 'nebulae' in an effort to understand what they really were. He found, in study-

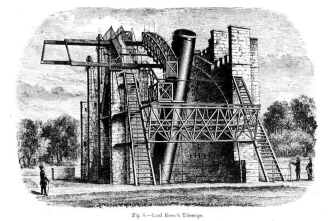

Fig. 6.—Lord Rosse's Telescope.

Lord Rosse's giant reflecting telescope, Birr Castle. It was with this instrument that Rosse studied nebulae and clusters.

ing the starry nebulae, that they were like gigantic 'catherine wheels' of stars. They had a spiral shape which he depicted in numerous fine drawings which he later published. He found that some were seen 'flat on', others tilted at various angles to the galactic plane, while some were actually seen 'edge on'—the view we get of the Milky Way.

Lord Rosse's work was extremely important, and Birr Castle remained a centre for both British and foreign visitors well after the great telescope fell into disuse in about 1880. Rosse had established there to be two distinct classes of 'nebulae': stellar and gaseous. However, it was unclear as to whether or not the former were part of our own galaxy. A decision on this question was not forthcoming until astrophotography was developed.

### Magellanic Clouds and the Period-luminosity Law

In 1912, at Harvard, the astronomer Henrietta Leavitt started to study photographs of what are called the Magellanic Clouds—two stellar concentrations which appear to be 'broken-off' parts of the Milky Way, and which lie quite close to the south celestial pole. These are out of sight of northern-based astronomers. At that time it was generally believed that both clouds were a part of our own galaxy. Miss Leavitt began by studying the Cepheid variables which lay within the clouds. After much diligent research she established that the Cepheids with the longer periods were always brighter than those with shorter ones. She had discovered what we now know as the Cepheid 'Period-luminosity Law'.

The significance of this discovery had far-reaching applications. Since it was known that the Magellanic Clouds were extremely distant, it must follow that, to all intents and purposes, any Cepheids lying within the Clouds must be considered equidistant from Earth.

It follows, then, that the brighter, longer-period variables must actually be more luminous than their

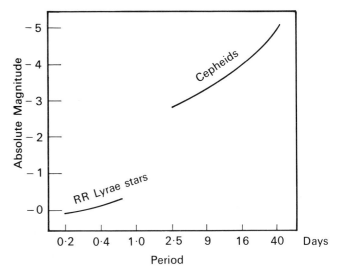

Period-luminosity curves for Cepheid and RR Lyrae variable stars. These stars have become valuable in measuring stellar distances.

shorter-period companions. The Law states, therefore, that the period of a true Cepheid is proportional to its real luminosity: once the period of such a variable is known, it becomes a relatively simple matter to calculate its distance.

As an example we can briefly look at two well-known stars of this type: the 'type' star, Delta Cephei and another Cepheid, Eta Aquilae. The latter star has a period of 70 days and its apparent brightness is the same as that of Delta Cephei. However, Eta has a longer period than Delta Cephei and so must be the more luminous of the two. It follows, therefore, that Eta Aquilae is the more remote of the two.

The Cepheids became valuable in measuring distances across the galaxy and, of course, in measuring the size of not only the galaxy, but of more distant objects. The RR Lyrae stars we have already discussed can best be considered as a special case of Cepheid behaviour, and we already have seen how their worth has been proved.

Henrietta Leavitt discovered the Period-luminosity Law but it was left to Edwin Hubble, another Ameri-

The dome of the great 100-in reflector of the Mount Wilson Observatory, California. Hubble measured the distances of both starry and gaseous nebulae with this instrument.

can worker, actually to measure the distances of both stellar and gaseous 'nebulae'. Using the great 250 cm (100-in) reflector at Mount Wilson, in California, Hubble discovered Cepheids within the Great 'Nebula' in Andromeda (M31) and was thus able to measure its distance. His results supported Herschel's shrewd guess about the 'starry nebulae': M31 lay at the immense distance of 2·2 million light years, well beyond the confines of our own galaxy. In fact, as later spectroscopic work has shown, it is another complete galaxy in its own right. Although it seems incredibly remote, it is actually one of our nearer galactic neighbours. The two Magellanic Clouds are also galaxies and both, incidentally, are slightly nearer to us than M31.

The starry 'nebulae' are thus not nebulae at all; they are galaxies. So we now speak of the Great Andromeda Galaxy, not the 'Nebula in Andromeda', as Messier did. What of the 'true' nebulae?

Early spectroscopic work revealed that these gaseous clouds are composed largely of illuminated gas and dust. Hydrogen is very abundant, but oxygen, nitrogen and helium feature too. The true nebulae actually give out light energy, and are not simply lit up by neighbouring stars. However, this is not to say that there are not stars within them. It is currently believed that the nebulae which, by the way, lie within our own galaxy, are the birthplaces of new stars. The brighter nebulae nearly always contain hot white stars which are known to be youthful stars on the cosmic scale. It is the presence of these young hot stars which helps to illuminate the gases surrounding them.

## Supernovae and Novae

Probably many readers will have heard of the Crab Nebula. It has been the subject of several television feature programmes during the past few years. It lies near the third magnitude star, Zeta Tauri, and

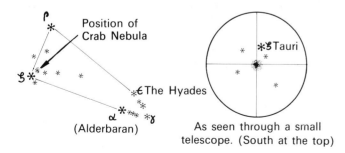

The position of the Crab Nebula, M1 Tauri. This represents the wreck of an old star which broke up in the year 1054.

represents the wreck of a star which exploded as long ago as the year 1054. The exploded star gave rise to what is termed a *supernova* and this was recorded at the time as a brilliant celestial object. At its brightest, the supernova shone with the light of 10 million suns, but gradually it faded and today only a faint White Dwarf can be seen within the nebula, even with the most powerful telescopes.

The Nebula itself is a typical galactic cloud of gas, and astronomers have found that it is still expanding at a tremendous rate. It is amazing to think that we are still seeing the remains of a celestial explosion that was observed by our ancestors nearly a thousand years ago.

Supernovae are relatively rare. However, less spectacular events, called simply *novae* are more common. As I write this book, there is a recorded nova in the constellation of Vulpecula. Here, among established and well-charted star fields, a stellar object of magnitude 6·0 has appeared. It is now visible in a small telescope and represents some kind of explosion out in the vastness of space, where a star has suddenly flared up. Like all novae, it will probably soon subside and sink back into obscurity along with its numerous predecessors.

To return once again to the galaxy: we know it is a vast spiral mass of stars and that our Sun lies along one of the arms. Associated with it are gas clouds, or nebulae, and larger globular clusters which from a kind of external ring to it. By everyday standards it is a huge system, and it extends over unbelievably large distances. Nevertheless, to astronomers it is relatively close, and they know of much larger star systems; each day new discoveries are being made. The universe seems to be endless. As new methods of studying the stars are developed so the frontiers of space are pushed further and further back. In the next chapter we shall take a look at some of the more distant objects astronomers have discovered, as well as at some closer but equally enigmatic objects.

# BEYOND THE GALAXY

### Hubble's Classification of Galaxies

We know that our own galaxy is a spiral structure. All such structures appear to be true galaxies but not all galaxies are spirals. Some years ago, Edwin Hubble distinguished various kinds of galaxy. Spirals similar to our own he classified as type S objects and he further distinguished between tightly coiled types (type Sa), moderately tight coils (Sb) and loosely coiled galaxies (Sc). Our own is of the latter class.

Other spirals have arms which seem to spring from bar-like bridges of stars which cross the galactic centres: these he called the barred spirals (type SB). Then again there were galaxies that were elliptical in form and appeared to have no spiral arms at all. These

The spiral galaxy NGC2309 in the constellation of Leo.

Hubble called type E objects and he subdivided the class on the eccentricity of the ellipses. Thus his class E0 galaxies are nearly circular, while class E7 are long and narrow. Lastly he described galaxies, like the Magellanic Clouds, which had no discernible regular form at all. These are his type I objects—irregular galaxies.

Let us return for a moment to the Magellanic Clouds. These have no proper shape, as Hubble recognised, and they lie at a distance of 180,000 light

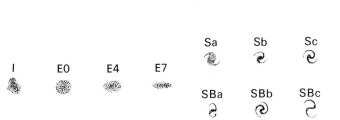

|  | E0 | E4 | E7 | Sa | Sb | Sc |
|---|---|---|---|---|---|---|
| I |  |  |  | SBa | SBb | SBc |

Hubble's classification of galaxies. In general terms he distinguished between spirals, barred-spirals and irregular types.

years. Within the larger cloud there is at least one fine nebula, far larger than the more famous one in Orion (M42). There are also nebulae within the more distant Andromeda galaxy (M31), as well as globular clusters. The stars within both galaxies are generally far more luminous than our own Sun, which, you will remember, is a yellow dwarf and a rather unspectacular star by most standards.

In researching the galaxies, Hubble and his colleague, Milton Humason, also found that they tended to cluster into groups. The Magellanic Clouds, M31 and the galaxy known as M33 Triangulum, together with our own galaxy, form what is known as the 'local group'. More distant galaxies appear to group in the same way, although many of the groups are far larger than our own. One particular group, located in the constellation Virgo, contains at least 1,000 galaxies and lies fifteen million light years away.

A cluster of galaxies on the constellation of Coma Berenices. These star systems lie at the immense distance of 200 million light years and consists of about 1,000 galaxies. This photograph was taken with the 200-in reflector at Mount Palomar, California.

The same two workers next turned their attention to how the various galaxies and cluster of galaxies were moving. This they did by using the spectroscope. Light, being a form of electromagnetic radiation, vibrates at different wavelengths. Blue light

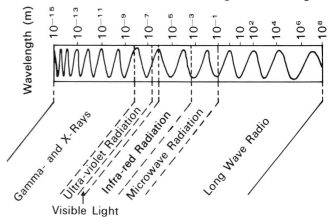

The electromagnetic spectrum. Long-wavelength radiation (infrared) is invisible to our eyes but can be recorded. Radio waves are of even longer wavelength. X-rays are of very short wavelength.

vibrates at shorter wavelengths than red. Radiations of longer wavelengths than red light are invisible to our eyes and are called infrared vibrations. At longer wavelengths again we pass into the realm of radio waves.

Because light behaves in many ways like sound waves, we can draw here a useful analogy, much hackneyed but still valuable. If one stands on a railway platform as an express approaches, the pitch of its whistle rises as the train approaches; once it passes and recedes the whistle pitch decreases. The wavelength of the signal has audibly lengthened as the train moves away.

Light behaves in similar fashion. If a bright object is receding from us, the dark lines within its spectrum are shifted towards the red (longer wavelength) end of the band. This is known as the 'Doppler Effect' after its discoverer, and also as the 'Red Shift'.

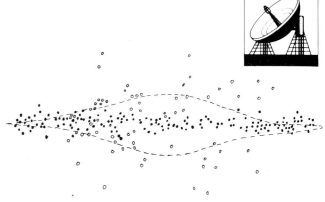

## Hubble's Law

Hubble found that members of the 'local group' were effectively stationary with respect to one another. However, the more distant galaxies showed marked Red Shifts indicating that they were moving away at great velocities. In fact, it appeared that the more distant a galaxy, the faster was its velocity of recession. This simple relationship became known as Hubble's Law.

If Hubble's Law is true, and there is little reason to doubt its validity, then the universe is steadily expanding all the time. The further away a galaxy or cluster of galaxies is, the faster it appears to be travelling. Some of the most distant objects appear to be travelling at almost half the speed of light. However, there is no fear that they will shortly disappear from our view, the scale of the universe being so immense that their withdrawal is not apparent from century to century.

It has been possible to learn at least something of the development of a galaxy by studying the distribution of stars, nebulae and dust within it. Thus the spiral arms of M31, the Great Andromeda Galaxy, are dominantly composed of hot, white stars. In the central regions however, the majority of the stars are large Red Giants.

## Baade's 'Populations'

This fact was initially realised by the eminent astrophysicist, Walter Baade, who made a series of important observations during the Second World War. Baade discovered that there seemed to be two distinct types of star 'population'. In galactic arms were concentrated hot, white, stars of spectral classes A and B, together with many nebulae and interstellar dust. In the centre of the galaxies, however, were Red Giant stars, while there were few, if any, nebulae and no dust. Baade called the stars and nebulae of the spiral arms population I, while those of the galactic cores he called population II.

The distribution of stellar populations within the galaxy. New stars are being born in the galactic arms while the central regions contain rather old stars.

As I mentioned earlier, the white stars of population I are young while the red stars of population II are older and well through the major part of their life. Thus we believe that along the galactic arms, new stars are being born out of the clouds of gas and dust that exist there. Towards the central regions, on the other hand, all the gas and dust has already been used up, so that the stars there are much older.

## Radio Telescopes

All that we have so far considered has been the result of visual or photographic observation. In 1931 a new discovery led to means of studying the universe by radio waves. During that year, Karl Jansky, an expert in radio technology, found that an annoying background 'hiss' that had been blemishing signals he had been receiving was emanating from none other than the Milky Way. This was the start of radio astronomy, and its development revolutionised modern cosmology. Jansky himself never followed his findings any further; this was left to others.

After the Second World War ended, radio 'telescopes' were constructed at various sites in order that the radio waves could be received and their positions located. Such instruments are in reality very powerful radio receivers. The famous Jodrell Bank instrument is of the 'dish' type, but others use arrays of aerials sited along the ground. Both types are aimed at receiving radio waves from distant sources and con-

The dish-type radio telescope of the University of Manchester, at Jodrell Bank, Cheshire.

centrating them sufficiently for their behaviour to be studied. No radio telescope provides a visual image, the wave patterns being recorded on drums by pen recorders and then analysed in the laboratory.

Very soon after the first good telescopes had been built, it became clear that the Sun was a source of radio waves. Later the planet Jupiter was found to send out similar vibrations, but more exciting still was the location of a strong source of radio waves within the region of the Milky Way, and centred on M1 Tauri, the Crab Nebula.

The Crab, as you will remember, is a supernova remnant, first recorded as a brilliant daytime star by the Chinese in 1054 AD. Lord Rosse drew it through his greater reflector at Birr Castle and subsequently photographs were obtained with some of the world's largest telescopes. It was known from visual observations that the gases within it were expanding outwards

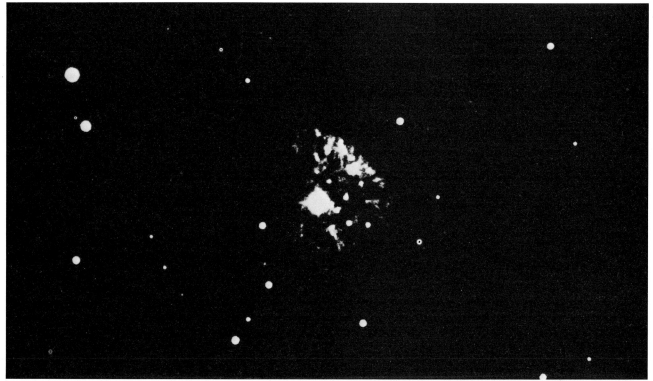

The 'Crab' Nebula. The remnant of a supernova explosion. Photographed with the 200-in Palomar reflector.

from the old explosion centre at a tremendous rate. It turned out to be one of the strongest emitters of radio waves known. It also sends out X-rays—electromagnetic radiations of even greater wavelength than radio waves; thus it is active in all parts of the electromagnetic spectrum.

Other supernovae are known, not many it must be admitted, but both the one in Cassiopeia which flared up in 1572 and was described by none other than Tycho Brahe, and that in Ophiuchus described by Kepler and active in 1604, have also been found to be strong emitters of radio waves. Neither of the latter has a visible mantle of expanding gas, like the Crab, but both clearly are active.

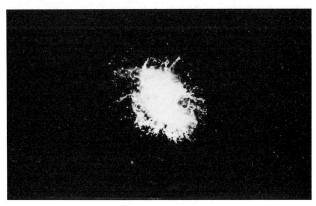

Expanding nebulosity around Nova Persei (1901).

One of the drawbacks of early radio telescopes was that they did not pinpoint the sources of radiation in the same way as do optical telescopes, and refinements had to be made so that the actual sources of the radiations could be tracked down. In the 1960s researchers applied themselves to solving this problem and in doing so came up with one of the more astonishing discoveries of the century.

## Pulsars, White Dwarfs and Neutron Stars

Jocelyn Bell, an astronomer working out of Cambridge, made a startling discovery. She seemed to be receiving rapidly pulsating signals from an unknown source. These had a very short period, rather like the emissions from a radio beacon. Initially it was thought the reception must be from a terrestrial source, but subsequently it was shown quite clearly that these waves came from outer space. At one time the bemused Cambridge astronomers secretly entertained the amazing idea that the pulsating signals might be from an alien civilisation which was trying to contact us across the vastness of space. However, this idea was later shelved.

Was this pulsating object, christened a *pulsar*, unique? A search was immediately instigated and very soon it became clear that it was by no means unique. At the present time, over a hundred such sources are known. Each pulsar lies within our own galaxy and despite careful searching very few can be detected optically, even with the world's most powerful telescopes. What could they be?

Two theories were favoured at first, pulsars could either be: (1) vibrating White Dwarfs, or (2) *neutron stars*. White Dwarfs are small, dense stars, as we saw in chapter 4, but although they are small compared to most other kinds of star, they still have diameters in excess of our own Moon. Such a body could not possibly spin or vibrate rapidly enough to produce the very short-period pulsating signals that were being received. This left neutron stars as a consideration.

While all this was being thought out, another piece of news arrived. Pulsar-like signals had been picked up from the Crab Nebula. This discovery immediately solved one query which had remained in astronomers' minds; namely, how long could such a pulsating source continue to radiate? Since the Crab 'exploded' over 900 years ago, this particular pulsar must have been pulsating for at least that long.

Eventually, after much painstaking work, the Crab pulsar was tracked down visually to a very faint, flashing object that was clearly not a White Dwarf. It had to be a neutron star.

When we were discussing the life history of a star, I showed how stars, like our Sun, evolved into Red Giants and finally contracted into dense White Dwarfs in their old age. However, not all stars are like the Sun and, in particular, stars which are more massive than the Sun, will behave in a rather different manner. As the fuel resources of very massive stars wane, the internal reactions get out of control and there is an inevitable massive explosion, producing a supernova event. Much of the stellar matter is blown away into space (as happened with the Crab), but the remaining material shrinks down into an incredibly dense body in which the protons and electrons are converted into neutrons, leaving us with a *neutron star*. These bankrupt stars are even more dense than White Dwarfs and it is impossible really to understand the state of matter in such a body. If I were to suggest that a thimbleful of neutron star material would weigh something of the order of ten thousand million tonnes, I should not be very wide of the mark. To me, at least, this is a very difficult thing to envisage.

A tiny body like a neutron star could quite easily spin rapidly enough to produce the observed, very rapid, pulsations of the radio observers. Astronomers believe they behave rather like beaming searchlights, in that only when we lie in the path of the beaming neutron star, do we receive the rapidly fluctuating signals. Pulsars, then, are among the weirdest objects in our galaxy, and represent truly bankrupt, senile stars that have suffered a catastrophic explosion such as that represented in M1, the Crab Nebula.

Radio astronomy allows us not only to study our own galaxy, but also helps in understanding how more distant objects behave. During the early days of radio observation it was soon found that radio emissions were being sent out by the extremely cold, rarified gas which existed between the stars themselves. As more telescopes were built, so increasing numbers of radio sources were located and catalogued. In particular there are hundreds of sources beyond the galaxy, many of which are exceedingly powerful radio emitters. For instance, the object M87 in Virgo is one of these very strong sources, while M82 in Ursa Major appears to send out radio waves as a result of cosmic explosions within it, possibly due to galactic collisions. One of the most remote radio galaxies so far discovered lies at the enormous distance of 5,000 million light years!

The distances of such galaxies are incredible and are of the same order of 'unimaginability' as the superdensity of pulsars. Once again, however, the universe had a surprise in store to equal, if not surpass this.

## Quasars

The majority of the extragalactic radio sources were clear spirals or brightish objects which, like the M82 galaxy, seemed to have undergone some internal disturbance. There were some objects, however, which seemed to correspond to very faint, bluish 'stars'. In the early 1960s, these odd objects were investigated vigorously. At Mount Palomar, Dr. Maarten Schmidt, armed with a series of radio observations made in Australia, looked again at the spectra of one of these 'stars' and to his amazement, found that the spectral lines were shifted so far towards the red end of the spectrum, that it must lie at a distance of thousands of millions of light years. It could not possibly be a star, since the largest conceivable stellar object would not be visible across such immense distances. Because it looked stellar, it was initially called a

'quasi-stellar object', and was later condensed into 'quasar'.

Since that time many more *quasars* have been identified. Several hundreds have now been added to the catalogues. Yet even today, with all our advanced equipment and knowhow, it must be admitted we are still unsure of what quasars really are. If their Red Shifts are due to Doppler Effect, and not to some as yet unexplained gravitational effect, then they are more remote than even the farthest known galaxies. The nearest must be at least 5,000 million light years away. Furthermore, their recession velocities are immense—in some cases speeds of almost ninety per cent of the velocity of light itself are indicated.

Such a body is almost unbelievable, for while it is perhaps no more than a few light years across, it must radiate energy equivalent to 200 galaxies thrown together. This appears quite absurd, but the observations seem to suggest that this is what we are dealing with. Quasars are an enigma. Clearly they are immensely significant, but it remains for astronomers to understand just what this significance is.

Space is vast. It holds many enigmatic objects within its confines, if indeed it even has these. At the present time astronomers are debating an even stranger concept which we can discuss as we near the conclusion of our journey.

Neutron stars—pulsars—are the dying remnants of stars, between one-and-a-half and ten times more massive than the Sun, which have spent their energies and which are left to exist as superdense, feeble 'glow-worms' in space. If an imaginary space probe landed there, it would find it quite impossible to get away again, as the escape velocity of such an object would be thousands of km per second—far greater than the velocities achieved by any rocket.

## Black Holes

There are, within the universe, even more massive stars in which the gravitational collapse that occurs during their middle age is so strong that they do not even become Red Giants. They do not become supernovae either, like the Crab Nebula. Under the increasingly severe gravitational collapse, they get smaller and smaller, denser and denser until their escape velocities become so vast that even light itself cannot escape from them. They have reached a point of no return. Nothing can travel faster than light, so we are left with what is called a 'Black Hole'.

Such an extreme type of gravitational collapse appears to be the fate of the more massive stars within our galaxy, and presumably, within any galaxy. Studies of X-rays emanating from known celestial objects have shown at least one candidate which may fit as a true and 'observable' Black Hole. This lies in the constellation of Cygnus and is known as X-ray source, Cygnus X-1. The visible component is a very luminous blue supergiant star, but it has an invisible companion of immense density. It appears that the invisible object has a mass equivalent to about fifteen suns. Very strong X-ray radiation is associated with this object, and at the present time bemused astronomers favour the idea that this is due to very hot gases which are spiralling into the vortex of the Black Hole.

We cannot delve deeply into the Black Hole concept in this short book. Belief in it arises, of course, from Einstein's 'Theory of Relativity' which explains how gravity operates and how both matter and light are affected by it. Astronomers are at present bemused but tremendously excited by the Black Hole idea. A Black Hole cannot be destroyed! It must devour matter continuously, sucking in more and more all the time, and growing larger as a result.

We shall never be able to 'see' a Black Hole. The only way of asserting it exists is to study the gravitational effects experienced by nearby objects. This is why binary stars, in particular those with invisible, superdense components, have become amongst the most observed celestial objects. When one thinks about this, one realises that it is not too far removed in principle from the way in which astronomers of the nineteenth century found the planet Neptune.

Black Holes and quasars must remain the domain of the professional astrophysicist. The layman and amateur scientist may follow the arguments and derive enjoyment from seeing how the tide of opinion turns as the years pass by. In practical terms, however, the heavens are full of beautiful and fascinating objects one can *see*. The appendixes which conclude this book are meant for those of my readers who now feel armed with a basic knowledge which will enable them to derive many hours of enjoyment from looking at the sky and wondering at what is there!

# EPILOGUE

Before this book finally reaches my readers further advances will have been made in various fields of astronomy. At the time of writing Pioneer 10 wends its lonely way out of the solar system, while the probe, Pioneer 11, is still on its way to rendezvous with the planet Saturn. Astronomers wait with some excitement for the first close-up photographs of the planet's beautiful ring system. In the summer of 1977 the Americans launched another probe, one of the Voyager series, whose mission is for a more sophisticated appraisal of Saturn and the outer planets.

At the beginning of 1977, John Hosty, an amateur observer from Huddersfield, using half a pair of binoculars mounted on a tripod, discovered the first nova of the year, Nova Sagittae (Hosty), which was of magnitude 7·2 on the evening of 7 January, and its existence has since been confirmed by another observer. Whether it will increase in brightness or slowly fade again into insignificance remains to be seen. What is clear, however, is that the enthusiasm of the amateur astronomer, often using meagre resources, is an incentive to the professional to be on his toes.

The study of novae, like the Sagitta object and the Cygnus one discovered by a Japanese astronomer during the latter part of 1975, is of great importance in our understanding of a star's behaviour during different phases of its life. Amateurs and professionals alike will continue to be alert to the possibilities of finding new novae in the future.

### Invisible Astronomy

Another recent development has been in the field of 'invisible astronomy'. This is the study of radiations beyond the visible end of the electromagnetic spectrum.

It is over 150 years since Sir William Herschel stated that the Sun emitted radiations in the infrared. However, it is only very recently that astronomers have perfected their techniques to a level sufficient to enable them to receive and analyse infrared radiations from space. In so doing, they opened up a whole new facet of the science.

Infrared 'observation' allows us to detect materials which are at much lower temperatures than normal stars. The cooler an object, in fact, the further into the infrared will it emit. What the new technique allows us to study is interstellar dust.

Interstellar dust radiates only when it is heated by nearby stars. Thus, when large scale infrared emissions are received from a 'dark' region of space it implies that stellar sources exist there, but are invisible to us due to the presence of intervening clouds of dust.

One of the first objects to be studied in this way was M42—the Great Nebula in Orion. Hitherto this had been considered a relatively small, tenuous cloud of hydrogen gas in the ionised state. Infrared studies changed this idea. For a start, a number of distinct infrared sources were found to exist in the vicinity and elsewhere in the Orion region. The fact that these cannot be located visually, means that the extent of the dust cloud is far greater than had been anticipated. In fact it is now clear than M42 is far larger, denser and more massive than anyone had expected. It seems that had the stars of the 'Trapezium' not managed to burn their way out of the cloud, we should not even have known of its existence, for it is their visible radiation which allows us to see M42 in the familiar way we do.

Similar gas and dust clouds, we know, are found in the spiral arms of our own and other galaxies. Infrared studies are now being made of these population II regions in an effort to understand them better and improve our picture of galactic structure and evolution.

Visible astronomy—the classical form—has now been supplemented by 'invisible' astronomy. Instruments receive stellar and galactic messages from radio, X-ray, and infrared sources. Together the old and the new astronomies slowly increase our knowledge of the universe.

# Appendix 1
# Some Interesting objects for the Naked Eye, Binoculars or a small Telescope

I have selected a small number of celestial objects which may be seen by the astronomer with only rudimentary optical aids. While this places restrictions on the list, the observer who is fortunate enough to view those subjects in it will, nevertheless, have seen some marvellous sights. Should my reader get bitten by the astronomy 'bug', I would refer him or her to: *Norton's Star Atlas, Yearbook of Astronomy, Naked Eye Astronomy* or the *Handbook of the BAA* for further information.

| Object | R.A. h m | Dec. ° | Comments |
|---|---|---|---|
| M31 Andromedae | 00.40 | +41 | The Great Andromeda Galaxy. Visible to the naked eye |
| R Böotis | 14.35 | +27 | A variable star which just precedes ε Böotis. At maximum the star reaches 6th magnitude, but it may decline to as low as 12th magnitude. The period is 223 days, for half this period it is visible in binoculars |
| M44 Cancris | 8.37 | +20 | The *Praesepe* Cluster. A very large cluster between γ and δ. It is quite easy to spot with the naked eye and is a fine sight in binoculars |
| M3 Canes Venatici | 13.40 | +29 | A brilliant globular cluster. It lies almost half way between *Arcturus* and α Canes Venatici. Small telescope |
| M41 Canis Majoris | 6.45 | −21 | A fine loose cluster which is just visible to the naked eye. It lies about 4° south of Sirius |
| 1 and 2 Capricornis | 20.15 | −13 | A naked eye double star. The components are of 3rd and 4th magnitude and separated by 376″ |
| β Capricorni | 20.18 | −15 | Close to α Capricorni. A yellow 2½ magnitude star with a blue 6th magnitude companion, 205″ apart |
| δ Cephei | 22·27 | +58 | Reddish star. It varies from magnitude 3 to magnitude 4½ in five days. It also has a companion of 7th magnitude at 41″. Binocular object |
| o Ceti | 2·17 | −03 | *Mira.* Very red star. Varies from magnitude 2 to 10 |
| 32 and 33 Comae Berenicis | 12·50 | +17 | A binocular double star. The components are of magnitude 5½ and 6, separated by 195″ |
| M53 Comae Berenicis | 13·11 | +18 | A fine globular cluster just visible to the naked eye. Closely follows α Comae Berenicis about ½° north of it |
| β Cygni | 19·29 | +28 | One of the most beautiful double stars. A golden 3rd magnitude star has a 5th magnitude blue companion. Separation is 35″ |
| 16 and 17 Draconis | 17·31 | +55 | Extend a line from γ and β Draconis, for twice the distance. Binoculars will reveal a 5th magnitude double star |
| M35 Geminorum | 6·06 | +24 | A fine loose cluster. Just visible to the naked eye; a fine sight in binoculars |
| ζ Geminorum | 7·01 | +21 | A 4th magnitude variable star with a 7th magnitude companion at 96″ distance. Binoculars |

| Object | R.A. | Dec. | Comments |
|---|---|---|---|
| M13 Herculis | 16·40 | +37 | The Great Globular Cluster in Hercules—the finest in the northern heavens. It is visible to the naked-eye, about one-third the distance from η to ζ Herculis |
| M92 Herculis | 17·16 | +43 | A fine globular cluster, smaller but brighter than the above. It lies at the apex of a tringle with π and η Herculis |
| τ Leonis | 11·25 | +03 | A wide double star. The stars are of 5th and 7th magnitude, separated by 92″. Binoculars |
| α Librae | 14.48 | −16 | A wide double star. It lies almost on the ecliptic. The stars are of 3rd and 6th magnitude, separated by 231″ |
| ζ Lyrae | 18·43 | +38 | A fine double star. The components are of magnitude 4 and 5½, separated by 44″ |
| ε Lyrae | 18·43 | +40 | A 'double double' star. The main stars are of 2nd and 5th magnitude. These are visible in binoculars. A small telescope reveals that each, in turn, is itself double |
| NGC2232 Monocerotis | 6·26 | −05 | An open cluster formed around 10 Monocerotis. It is visible to keen-sighted persons but is better seen in binoculars |
| β Orionis | 5·12 | −08 | The 1st magnitude star has a 7th magnitude companion. The two are 10″ apart. This is a standard test for a 2″ O.G. |
| α Orionis | 5·53 | +07 | *Betelgeuse.* An orange-red, irregular variable star well worth watching |
| M42 Orionis | 5·33 | −05 | The *Great Nebula in Orion.* The star θ Orionis is within it. This forms what is called the 'Trapezium' of four stars from 4th to 6th magnitude |
| χ Persei | 2·17 | +57 | A line from γ to δ Cassiopeia, produced for twice its length brings one to what at first appears to be a hazy patch. In binoculars this is seen to be a double cluster of very grand proportions. It is known as the '*Sword Handle*' |
| β Persei | 3·05 | +41 | *Algol*—The 'Winking Demon'. See text |
| *Pleiades* | 3·44 | +24 | The famous open cluster, also known as the 'Seven Sisters'. Try charting the group, using binoculars |
| M8 Sagittariis | 18·10 | −24 | An open cluster of stars which includes considerable nebulosity. Hence its inclusion in Messier's catalogue |
| θ Tauri | 4·36 | +16 | A double star which may be seen with the naked eye. Their components are of the 4th and 4½ magnitude and are 337″ apart. They form part of the *Hyades* cluster |
| ζ Ursae Majoris | 13·22 | +55 | *Mizar* (magnitude 2), and Alcor (magnitude 5) form a naked eye pair. A small telescope shows that Mizar has a companion at 15″ |

# Appendix 2
# Table of Planetary Elements

| | Mercury | Venus | Earth | Mars | Jupiter | Saturn | Uranus | Neptune | Pluto |
|---|---|---|---|---|---|---|---|---|---|
| Mean distance from Sun: | | | | | | | | | |
| Millions of miles | 36·0 | 67·2 | 92·9 | 141·5 | 483·3 | 886·1 | 1,783·0 | 2,793·0 | 3,666·0 |
| Millions of km | 57·9 | 108·1 | 149·5 | 227·7 | 777·6 | 1,425·7 | 2,868·8 | 4,493·9 | 5,898·6 |
| Sidereal period | 88 days | 224·7 days | 365·3 days | 686·9 days | 11·86 years | 29·46 years | 84·02 years | 164·79 years | 249·7 years |
| Period of axial rotation | 58·6 days | 243·1 days | 23 h 56 m | 24 h 37 m | 9 h 50 m | 10 h 14 m | 10 h 45 m | 15 h 50 m | 6·39 days |
| Orbital inclination of ecliptic | 7° 0′ | 3° 23′ | 0 | 1° 51′ | 1° 18′ | 2° 29′ | 0° 46′ | 1° 46′ | 17° 6′ |
| Inclination of Planet's equator to orbital plane | 0 | 2·2° | 23·5° | 25·2° | 3·1° | 26·7° | 98·0° | 29·0° | ? |
| Equatorial diameter: | | | | | | | | | |
| miles | 2,900 | 7,700 | 7,929 | 4,200 | 88,700 | 75,100 | 29,300 | 27,700 | ?3,600 |
| km | 4,666 | 12,389 | 12,755 | 6,758 | 142,700 | 120,800 | 44,600 | 50,600 | ?5,792 |
| Mass (Earth=1·0) | 0·05 | 0·81 | 1·00 | 0·11 | 318 | 95 | 15 | 17 | 0·18 |
| Volume (Earth=1·0) | 0·06 | 0·88 | 1·00 | 0·15 | 1312 | 763 | 50 | 43 | ? |
| Escape velocity: | | | | | | | | | |
| mps | 2·6 | 6·4 | 7·0 | 3·2 | 37·1 | 22·0 | 13·9 | 15·5 | ? |
| kps | 4·2 | 10·3 | 11·2 | 5·1 | 59·7 | 35·4 | 22·4 | 24·9 | ? |
| Mean density | 5·5 | 5·2 | 5·5 | 3·9 | 1·3 | 0·7 | 1·6 | 1·5 | ~6·0 |

# Appendix 3
# Books about Astronomy

There is, of course, a vast literature about astronomy and astronomical matters. In this short appendix, I have listed a relatively small number of quite recent books, which I think may be of interest both to the layman and to the school student who is starting to become interested in the subject.

## General Astronomy

*Star and Planet Spotting.* P. Lancaster Brown (Blandford)
*The Discovery of our Galaxy.* Charles A. Whitney (Angus and Robertson)
*New Frontiers in Astronomy.* Reprints from 'Scientific American' edited by Owen Gingerish (Freeman and Co)
*Infrared—the New Astronomy.* David A. Allen (Keith Reid Ltd.)
*The Sky at Night, 5.* Patrick Moore (BBC Publications)
*Guide to the Stars.* Patrick Moore (Lutterworth)
*Black Holes in Space.* Patrick Moore and Iain Nicholson (Ocean Books—paperback)
*All About Space Exploration.* Peter Cattermole (Carousel Books—paperback)
*The Amateur Astronomer.* Patrick Moore (Lutterworth)

## Solar System and Moon

*The Solar System.* Reprints collected from 'Scientific American' in 1975 (Freeman and Co)
*The Sun and the Amateur Astronomer.* W. H. Baxter (Lutterworth)
*The Comets.* Patrick Moore (Keith Reid Ltd.—paperback)
*Comets.* Russel Ash and Ian Grant (Universal-Tandem Publishing Co.—paperback)
*Lunar Science—A Post-Apollo View.* Stuart Ross Taylor (Pergamon—paperback and hardback)
*Guide to the Moon.* Patrick Moore (Lutterworth)

## Historical Astronomy

*Royal Greenwich Observatory.* W. H. McCrea (HMSO paperback)
*Astronomy—A Popular History.* Johann Dorschner *et al.* (Almarck)
*Watchers of the Stars.* Patrick Moore (Michael Joseph)
*Galileo.* Colin Ronan (Weidenfeld and Nicolson)

## Telescopes

*The Young Astronomer and his Telescope.* Patrick Moore (Keith Reid Ltd.—paperback and hardback)

## Star Atlases and Handbooks

*Colour Star Atlas.* Patrick Moore (Lutterworth)
*Norton's Star Atlas.* Gall and Inglis (the sixteenth edition of this famous work)
*1979 Yearbook of Astronomy.* Patrick Moore (ed.) (Eyre and Spottiswoode)
*Naked-Eye Astronomy.* Patrick Moore (Lutterworth)
*Chart of the Stars* (Geo. Philip and Son)
*Philip's Moon Map* (Geo. Philip and Son)
*Philip's Planisphere.* 13 cm (5 in) diameter; 25 cm (10 in) diameter (Geo. Philip and Son)

# Appendix 4
# Astronomical Societies and Journals

The advantages of joining an astronomical society are numerous. One can meet people of like interest, hear lectures given by experts in various astronomical fields, and obtain advice on purchasing or building telescopes of various kinds. Some societies own their own telescopes which are made available to members for observations. The larger bodies also house libraries in their premises, from which a multitude of books may be borrowed and rarer volumes consulted. These also publish regular journals which are a mine of information, and to which the keen amateur can contribute articles and observations.

There are three amateur organisations of national standing. These are:

*The British Astronomical Association*
Secretary: J. L. White, Burlington House, Piccadilly, London, W1. Meetings held monthly in the above premises, on the last Wednesday of each month between October and June. The Association publishes a very good Journal.

*Junior Astronomical Society*
Secretary: 58 Vaughan Gardens, Ilford, Essex. Meetings are held on the last Saturday of January, April, July and October, at 2.30 pm at Alliance Hall, Palmer Street, London, W1. This is specifically run for the beginner and younger enthusiast. Subsequently, he or she would graduate to the British Astronomical Association, described above.

*Irish Astronomical Society*
Secretary: D. Beesley, The Planetarium, Armagh, Northern Ireland. Meetings are held fortnightly in Belfast and monthly in Armagh.

Similar, but smaller organisations exist in a large number of towns in the United Kingdom. There is not room here to list all the addresses of these. The interested astronomer should contact their local library or information office, or can consult the list given in the 1977 *Yearbook of Astronomy* Patrick Moore (ed.), and listed in appendix 3.

Societies exist in the following towns:

| | | | |
|---|---|---|---|
| Aberdeen | Bristol | Cleethorpes | Farnham |
| Alloa | Burnley | Coventry | Glasgow |
| Altrincham | Carlisle | Crawley (Sussex) | Great Yarmouth |
| Aylesbury | Camberley | Crayford | Guildford |
| Bangor (Gwynedd) | Cambridge | Croydon | Haringey |
| Barrow-in-Furness | Cardiff | Dartington | Harlow |
| Birmingham | Chelmsford | Denbigh | Harrow (Middlesex) |
| Blackburn | Chelmsley | Derby | Haywards Heath |
| Bognor Regis | Chester | Dundee | Ilfracombe |
| Bradford | Chesterfield | Eastbourne | Isle of Wight |
| Bridgwater | Churchdown (Glos.) | Edinburgh | Leeds |
| Brighton | Clacton | Ewell | Leicester |

| | | | |
|---|---|---|---|
| Lincoln | Newcastle-upon-Tyne | Portsmouth | Southampton |
| Liverpool | Norwich | Preston | Southport |
| Livingston | Nottingham | Rayleigh | South Shields |
| (W. Lothian) | Oxshott | Ramsgate | Stoke-on-Trent |
| Loughton | Paisley | Reading | Swansea |
| Luton | Palmers Green | Rickmansworth | Thurso |
| Lytham St Annes | (London, N13) | Salford | Torbay |
| Maidenhead | Peterborough | Salisbury | Waltham Forest |
| Manchester | Plymouth | Scarborough | Warrington |
| Mansfield | Pontefract | Sheffield | Wolverhampton |
| Milton Keynes | Poole | Slough | York |

# INDEX